FOOD AND BEVERAGE CONSUMPTION AND HEALTH

CASSAVA

PRODUCTION, NUTRITIONAL PROPERTIES AND HEALTH EFFECTS

FOOD AND BEVERAGE CONSUMPTION AND HEALTH

Additional books in this series can be found on Nova's website under the Series tab.

Additional e-books in this series can be found on Nova's website under the e-book tab.

FOOD AND BEVERAGE CONSUMPTION AND HEALTH

CASSAVA

PRODUCTION, NUTRITIONAL PROPERTIES AND HEALTH EFFECTS

FRANCIS P. MOLINARI
EDITOR

nova publishers
New York

For permission to use material from this book please contact us:
Telephone 631-231-7269; Fax 631-231-8175
Web Site: http://www.novapublishers.com

Library of Congress Cataloging-in-Publication Data

ISBN: 978-1-63321-031-8

LCCN: 2014939983

Published by Nova Science Publishers, Inc. † New York

CONTENTS

PREFACE

Cassava is the most important of the root crops in the tropics and ranks fourth after rice, sugarcane and maize as a source of calorie for human needs. Since its introduction into Africa in the 16th century, cassava has spread throughout sub-Saharan Africa to become one of the dominant starchy staples in the diet of the people. Cassava is a major staple for about half of the Nigerian population and nearly 50 million tonnes of fresh cassava is produced annually in the country. This quantity is enough to meet the calorie requirement of 50 million people at about 200 calories per day. Cassava is important not just as a subsistence or food security crop, but also as a main source of cash income for producing households. Its production and processing provide employment and income for the rural poor, especially women and children. This book discusses the environmental impact cassava may have; the functional and nutritional characterization of cassava flour; cassava bread; other food products developed from using cassava roots and its derivatives; and cassava varieties and their breeding status.

Chapter 1 - The adsorption phenomena were addressed in this chapter as the main mechanism to remove toxic metal species from water. This way, metal species will adsorb onto a specific surface, which can be a synthetic material surface such as silica, cellulose, titania or natural material as cassava husks. So, the present chapter will provide the reader another possible use of cassava. Despite its beneficial effects over health it may also be applied successfully in the removal of metals from water due to its chemical composition, which include nitrogen, oxygen and sulphur atoms. Chemical investigation was performed in order to identify organic groups, which are able to complex metal species in solution, and a detailed material preparation procedure was described. The batch experiments described may be applied to any other material. Although the material may be easily applied due to the simplicity of the technique, household applications are not encouraged once several chemical analysis are necessary prior to its safe use and to attest its good functioning.

Chapter 2 - Since, cassava root in its composition have not proteins forming gluten, the functional properties of cassava flour are not sufficient for elaboration of dough for baked goods, bread and pasta. However, if this raw flour is thermal and enzymatically modified, its properties could be adequate in order to develop different products. The goals of the study were to evaluate the nutritional and functional profiles of cassava flours that were elaborated from the edible portion of five clones of cassava a germplasm bank: Three sweets (≤50ppm cyanide) and two bitters (>50 ppm cyanide), and modify one of this flour by using thermal and enzymatic treatments, with and without addition of stearoyl-2-lactylate (SSL). Cassava flour produced by dehydration (60 °C x 24 hours) was modified: by boiling and autoclaving, in aqueous solutions of flours of 10% (with 2% NaCl), and 33 %, respectively, and dextrinizated with Termamyl®. The results showed differences in crude protein and ash contents in the different varieties of cassava flours assayed. They are also showed that the yield was less for the bitter variety than in the sweet cassava variety. The contents of moisture, crude protein, crude fat, ash, and dietary fiber was not function of the type of roots, varying from 8.1 to 10.6%; 1.3 to 3.5%, 0.4 to 1.1%, 2.0 to 2.6%, and 2.3 to 6.1%, for sweet and bitter varieties, respectively. Carbohydrate content was similar in all flour, exceeding the 80% in all of them. The color profile has also varies with cassava variety. Data are showing that the use of SSL, NaCl and the dextrinization treatment affect the moisture and ash contents significantly ($p \leq 0.05$). The use of SSL decreased the dehydration effect on the flours during the drying; the NaCl use and dextrinization treatment increased the ash content. The flours modified thermically and enzimatically were darker than the raw flour. The particle sizes were also affected by the treatment and SSL addition, showing a granular size varying from 125 to over 250 microns. The combination of dextrinization and SSL increased the granular size beyond 125 microns. Except for the dextrinizated flours (with and without SSL), which have not developed an amylographic profile, the addition of SSL have also decreased the overall viscosity, measured by using the amylograph. The boiling and retorting treatment have produced a type IV farinogram-curve. The SSL addition changed drastically this curves to those of type I, and the enzymatic treatment to the type II. These flours could be used for baked goods manufacture and could also be blended with type IV flour to the bread making.

The NaCl addition showed poor effects in the farinographic profile. The modified flours showed total granular disaggregation when were analyzed by SEM, indicative of total gelatinization. As conclusion, it suggest the feasibility to use these flours for the production of numerous baked goods, bread, and pastas gluten-free, since, they showed a wide spectrum of nutritional and functional properties by effect of additive and the treatments applied.

Chapter 3 - Native cassava starch have adequate functional properties; for example, their swollen quickly to low temperatures producing clear, sticky and fibrous pastes, that make them useful for their use in different industrial applications. As a contrast, they also have some limitations in their properties; such as, low stability to high temperatures (90°C), to pH changes, and to mechanical process, that often are restrictive their industrial use. These limitations in its functional properties are improved and optimized by the modification chemical and/or physical of this starch. The goals of this study were to evaluate the proximate composition, physical characteristics, morphometric parameters and functional properties of starches isolated and purified at the laboratory from cassava clones of the Venezuelan Germplasm Bank, and those from commercial cassava starches (native and crosslinked) acquired in the local market, and to compare all of these properties with those of its modified counterpart by using the following methodologies: physical (extrusion, dehydration in double drum, gamma irradiation and microwaves) and chemical (crosslinked, crosslinked-acetylated). It was also assayed, at a pilot level, critical parameters of the industrial process of crosslinked-acetylated modification in order to optimize it. Results have shown starches with high purity, and with a composition lightly affected by the modification treatments used, with the exception by those irradiated, where an increment of the contents of total and reducing sugar, and a diminution of its pHs were observed. Color of the starches was affected by the physical methods, making their less white. The contents of amylose apparent of the chemical modified starches have slightly augmented to 35.1-36.6%; as compared with its counterpart commercial native (34.13%), with a degrees of modification of: 0.013 (cross-linked), 0.035 (acetylated alone) and 0.064 (cross-linked-acetylated). The functional characteristics of starches were affected drastically, as function of the kind and severity of the treatments. The rheological behavior of native cassava starches is of a non-Newtonian pseudo plastic fluid; increasing its viscosity, as a function of the diminution of the temperature and with a pronounced viscosity peak, high granular breakdown, and retrogradation tendency. In the case of modified starches, its overall viscosities, granular breakdown, and retrogradation tendency are diminished with the chemical modification treatments used. Conversely, the extrusion treatments have shown the opposing effects. The drum-dried starch had developed an initial viscosity, without maximum peak and was very stable during the cooking. With the exception of the starch modified by dehydration (drum dried and extrusion); where the grain have shown a partial or total destruction of its granular structure, and loss of its Maltese cross, the other modification treatments are only affecting with a lightly augmentation of its sizes, keeping intact their structure, with its edges affluent defined, and forms three-dimensional, depressions corresponding to the *hilum*, smooth surface, and without the pore presence. The patterns of X-rays diffraction of the starches chemically modified do not vary, when comparing to its native counterpart. As conclusion it was demonstrated the feasibility to improve the cassava starch properties by modification for to extend its use in the different industries.

Chapter 4 - Food industry is interested in the study of sub-utilized plant species as a contribution to consumers' diet diversification and the incorporation of ingredients with specific technological properties.

In Argentina, more than 180,000 tonnes per year of cassava roots are harvested and this production supplies the domestic consumption and part of the local starch market. The Food and Agriculture Organization of the United Nations (FAO) has pointed out that many developing countries could strengthen their rural economy and increase farmers' income by converting relative low cost cassava feedstock in high-value starch. Compared to those obtained from other sources, cassava starch has more clear and viscous pastes and is stable in acidic food products. It can be used in pharmaceuticals and natural thermoplastic products.

In this chapter, data referred to cassava production, consumption and processing as well as some results obtained from the technological and nutritional characterization of cassava roots and their derived starch and flour are presented. Cassava can be considered a very good source of starch and might be used as a functional ingredient, particularly for the formulation of gluten free products.

Chapter 5 - Cassava (*Manihot esculenta Crantz*) is a tuber of the Family Euphorbiaceae of the genus Manihot originary from South America. Is a popular edible root since centuries in many parts of Africa, Asia and South America. Cassava is an important energy sources for tropical areas of the world, rich in calories but deficient: in protein, fats, minerals and vitamins and contains in low proportions, poliphenols, tannins and cyanogenic compounds.

Geographical localization of cassava bread supplier: Estado (Edo) Sucre, Venezuela east, "casaba Sucre", Edo Bolivar South of venezuela "casaba Amazon" and Edo Miranda, Venezuela Center "casaba Miranda". Using the enzymatic method of Prosky, was determined total dietary fiber(TDF), soluble dietary fiber(SDF) and insoluble dietary fiber (IDF) for "casaba Sucre" 5.4%, 0.8% and 4.6% respectively; "casabe Miranda" 5.2%, 0.7 and 4.5% respectively; "casabe Amazon" 6.3%, 1.6% and 4.7% respectively. Previous studies have reported average values for these varieties, IDF 4.92 - 5.67% and SDF 3.4 - 3.78% using the Asp method. The test for starch digestibility of "casabe" by pancreatic amylase showed to be very similar to the wheat starch thermally treated, however its rate of *in vitro* digestibility was relatively low 73%. Studies in the rat model (Sprague Dawley) using adult males, evaluated the effect of chronic consumption of "casabe" on the serum glucose concentration, weight, and length of the small intestine (SI). It was observed a reduction of glucose within 14 days of experiment, however, this level rose 16.3% above the control group at 21 days (end time). The group fed with a diet containing casabe: "casaba Sucre" or "casabe Amazon" showed a significant increase in wet weight and length of the SI, the highest values 24.2% and 8.94% respectively was shown for the group "casaba Sucre". The mechanism affected by the variation of glucose concentration, wet weight and length of the SI, is not yet known and still is in research. The varieties of "casaba Sucre", "casaba Miranda", and "casaba Amazon" were used to measure the effect on the serum concentration of triglyceride and total cholesterol. The group with the diet "casabe Amazon" showed the highest reduction value, 43% total cholesterol, 35% triglycerides; the variety "casaba Sucre" was 30-40% for total cholesterol and triglycerides. Also, the concentration of minerals showed a higher content of iron, copper and zinc compared with the control diet. The "casaba Sucre" showed the greatest iron and zinc content (6.49 and 36.76 ug/g diet respectively) and "casabe Amazon" showed the higher copper content 8.94 ug/g diet. A significant increase in the faecal excretion of iron and copper ($p<0.05$) was observed. The iron absorption increased 3.5% for "casaba Miranda", 6% for "casaba Sucre" and 32% copper for "casabe Amazon". The zinc absorption was less affected through the experimental time. The reduction of iron absorption for the "casabe Amazon" was associated to IDF, possibly due to the cationic interference process, suggested in the literature.

Chapter 6 - Starch is one the most common natural polymers used for the preparation of tablets by wet granulation. This technology employs a wet binder which provides to the excipient: drug mixture the proper binding characteristics to form granules able to be compressed into tablets. The aim of this study is to evaluate the viscosity, agglomeration behavior and tableting characteristics of cassava starch and to compare these properties to those of classic binders such as alginic acid, sodium alginate, sodium carboxymethyl cellulose, carrageenan, gelatin, polyvinyl pyrrolidone and gum arabic at 5 and 10% levels using ibuprofen as a model drug. Viscosity profiles were determined on a Brookfield viscometer. Wet binders were added to the microcrystalline cellulose-ibuprofen (60:40) mixtures on a ribbon blender and then granules were formed on an oscillating granulator. Granules were then compressed on a single punch tablet machine.

Results indicate that granule properties depended on the binder type, rather than on the binder level employed. Less plastic deforming materials such as alginic acid, gum arabic and PVP k30 rendered tablets with the best compactibility. However, no major differences were observed among the granule properties of cassava and corn starches.

Gum arabic and sodium carboxymethyl cellulose produced the most viscous dispersions (>50000 mPas), which in turn, showed no relation with the strength of the compacts produced. The largest particle size was found for granules made with alginic acid and gum arabic (~290 μm), whereas for most granules it ranged from 100 to 150 μm. The true density of the granules produced ranged from 1.31 to 1.40 g/cm^3. Granules having the largest size also presented the fastest flow rate (>16 g/s). Further, except for alginic acid, sodium alginate and cassava starch, bulk and tap densities ranged from 0.3 to 0.35 and 0.36 to 0.45 g/cm^3, respectively. Moreover, alginic acid and gum arabic produced granules with the lowest porosity (<65%). Likewise, corn starch was the most compressible material (>33%), whereas, alginic acid and gum arabic rendered the least compressible granules. Cassava starch presented the least elastic recovery and along with PVP k30 showed the best reworkability upon reprocessing.

This work demonstrated the potential use of cassava starch as a wet binder for the production of drug compacts by wet granulation.

Chapter 7 - The cassava plant is an important food item in many parts of the world. From this crop may be consumed its protein-rich leaves and its carbohydrate-rich roots with high starch content. However leaves and roots could be potentially toxic to the consumer due to the presence of cyanogenic glycosides, so they must be subjected to effective processing practices for reducing the cyanide in their derivative foodstuffs and these practices depend on the initial level of cyanide. It is of consensus that the roots from varieties called "sweets" have less than 50 ppm of cyanide, while the roots from the varieties called "bitters" have a higher cyanide level, above of 50 ppm. The roots from the "sweets" varieties are boiled or fried just enough to consume but the roots from "bitters" varieties must be crushed or soaked in water for a certain period of time to allow the cyanide volatilization. On the other hand, the fresh roots once harvested have a very short life and the post harvest losses are very high. Is why that the artisan ancient of the local natives and the local food industry have developed technologies of preservation and cyanide elimination, such as: flour elaborated with the "sweets" varieties, gari, farinha or casabe (cassava bread) elaborated with the "bitters" varieties, starch extracted from both varieties and that can be modified to elaborate a myriad of products. With the flour from "sweets" varieties, several products have been developed with or without mixing with wheat flour that have proved to be an alternative for consumers, especially those with dietary special regimes, obesity, cardiovascular disorders, celiac disease or phenylketonurics. Among the products made with 100 % cassava flour are: bread, pasta type "fettuccine", cake and pancake mixes and pizza. With the "catebía" (bitter cassava fermented flour) was prepared "casabe" with high resistant starch content. With the purified starch, native or modified, have been elaborated a kind of crackers and pudding flavored with chocolate. Increasing the consumption of these products, made using cassava roots, helping to reduce wheat flour dependence of the non-cereal producers countries and would be economically advantageous to reduce wheat import levels; this requires changing food habits and to inform the public about the benefits of consuming these products.

Chapter 8 - Cassava is a crop that has spread over a wide area of the American tropics from Venezuela to the northwest of Brazil. It have been established that roots containing less than 50 ppm of hydrogen cyanide are considered sweet, while those with cyanide above this value are bitter. However, it is unknown whether there are differences in the composition of the clones grown in the germplasm bank of the Venezuela country and information on the changes of composition as function of the harvest of cassava; which is between 7 and 11 months after the development of the plant. The goals were to characterize and to compare physical properties, edible portion, proximate composition and cyanide concentration of four varieties of cassava (*Manihot esculenta* Crantz) from the Germplasm Bank of Venezuela, in different stages of growth. Four varieties coded as: UCV-2192, UCV-2306, UCV-2593, and UCV-2681, were selected. Results of total cyanide classified the varieties UCV-2192 and UCV-2681 such as bitters and it was also observed that the total cyanide content were decreasing as plants was growing, varying from 360 to 118 ppm on the eleven month. The other clones (UCV-2306 and UCV-2593) shown less than 50 ppm of total cyanide and the concentration did not shown any tendency on the growing. The morphology of roots and plants are summarized as: roots and stems midsize well developed. According to these criteria, the sweet clones showed more homogeneous roots and medium size, better developed cylindrical with conical peduncles, than that those of the bitter one. The number of roots by plant was not affected by the cassava, increasing from 6 to 10 units at the last month. The weight of the roots increased significantly from 7 to the 11 months, varying from 169 to 410 g. However, the UCV-2681variety has roots with low weight (165-197g). In regards to the length and diameter of the roots, both parameters have increased with the stage of growing, and the bitter varieties have lower (16.7-17.6 cm) length than the sweet varieties (22.2-26.2 cm). The external color of the roots was lightly high in the bitter varieties, not varying with the growing stage. The edible portion and its color expressed with whiteness index were not affected in all varieties by the growing stage, varying from 70-72 % and 86.5-87.5% respectively. The edible portion has a tiny elevation in the sweet clones. The content of moisture (63.8-64.5%), crude protein (0.43-0.53%), crude fat (0.22-0.25%), dietary fiber (1.9-2.0%), ash (0.92-0.96%), and available carbohydrate (30.7-32.6%) were not affected by the growing stage. But, moisture and available carbohydrate contents were different with clone type: moisture content was less and available carbohydrate content higher in the sweet clones than in the bitter one. Consequently, the information obtained from this study of the cassava roots and as they are growing will assess their potential and will improve the products and by-products that are produced with them.

Chapter 9 - This chapter reviews the status and progress made so far in breeding for extra-early maturing cassava (*Manihot esculenta*) varieties. Late root bulking is a major factor leading to rejection and abandoning of improved cassava genotypes in sub-Saharan Africa. Extra-early bulking and high yielding (EB-HY) cassava varieties also referred to as extra-early maturing (EM) varieties in this review shorten the growth period from planting to harvesting. Some EM cassava genotypes have been selected for different agro-ecological zones in Nigeria after evaluation for two seasons at three locations representing three agro-ecological zones. Molecular markers associated with EB-HY in cassava were also identified using Bulked Segregant Analysis (BSA). The selected EM cassava genotypes offer cassava growers high productivity in less than 12 MAP. This will reduce the problems associated with long dry season experienced in most parts of cassava growing areas as farmers can afford to harvest their cassava plants at the beginning of dry season in areas where rainy season lasts for about 6-7 months. Also, the molecular markers identified for EB-HY in cassava avail breeders the opportunity to select for EM genotypes at early stage of breeding programme and thereby reduce the cost of such breeding programme.

In: Cassava
Editor: Francis P. Molinari

ISBN: 978-1-63321-031-8

Chapter 1

CASSAVA USE FOR ENVIRONMENTAL PURPOSE: REMOVAL OF METAL SPECIES FROM WATER

Gustavo Rocha de Castro, Alexandre de Oliveira Jorgetto, Marcos Henrique Pereira Wondracek and Adrielli Cristina Peres da Silva*
Instituto de Biociências de Botucatu-UNESP,
Dept. Química e Bioquímica, Botucatu, SP, Brazil

ABSTRACT

The adsorption phenomena were addressed in this chapter as the main mechanism to remove toxic metal species from water. This way, metal species will adsorb onto a specific surface, which can be a synthetic material surface such as silica, cellulose, titania or natural material as cassava husks. So, the present chapter will provide the reader another possible use of cassava. Despite its beneficial effects over health it may also be applied successfully in the removal of metals from water due to its chemical composition, which include nitrogen, oxygen and sulphur atoms. Chemical investigation was performed in order to identify organic groups, which are able to complex metal species in solution, and a

* E-mail: castrogr@ibb.unesp.br.

detailed material preparation procedure was described. The batch experiments described may be applied to any other material. Although the material may be easily applied due to the simplicity of the technique, household applications are not encouraged once several chemical analysis are necessary prior to its safe use and to attest its good functioning.

INTRODUCTION

Taking into account the growing Earth's population and the need to provide food and industrialized consume goods to supply the world demand, the use of sustainable or environmentally friendly practices are an immediate need in order not to compromise human life quality on Earth. Everyone who has access to media (internet, television or radio) has heard something about water, soil and atmosphere contamination and its consequences over our health. The contamination of environmental compartments is a consequence of non-sustainable practices which allows toxic substances (organic or inorganic) entering the environment at low or high concentrations. At high concentration a toxic substance may cause severe adverse health effects; however it does not mean that at low concentration deleterious effect does not occur, once the great majority of toxic substances such as pesticides and heavy metal species may accumulate in our organism. This way, in the present chapter the reader will find some environmental concepts regarding toxic metals species, their occurrence and fate, and how contaminated water can be purified or treated using synthetic and natural materials. Due to the chemical composition of cassava husks, it may be applied successfully in the purification of water, as we will discover in the following discussion of this chapter.

HEAVY METALS: OCCURRENCE, BEHAVIOR AND FATE

Heavy metal is an expression often used to refer to some species, located from the central part to the lower right-hand corner of periodic table that may pose risk to human health. The term "heavy" is probably due to its high density compared to other species and as examples we may mention copper, cadmium, nickel, mercury, lead and others. Among the mentioned species one or more metals may be also essential to the proper functioning of the body, as it is the case of copper, zinc, cobalt and others. As these species (so-called *essential element*) may also manifest toxicity at high concentration, the best way to refer to such species is by the term potentially toxic metals. On the other hand, toxic species are those that may cause deleterious effects over living organisms and does not present any known biological function.

Toxic and potentially toxic metals are usually found associated with other compounds in nature, such as organic matter, in the composition of rocks or adsorbed onto particles surface. During soil formation processes, rocks are undergone to several types of transformations due to physical and chemical phenomena, such as temperature variation, water condensation inside the pores, oxidation, and hydration or hydroxylation reactions. This releases metal elements by which rocks are composed and explains the natural occurrence of metal species (toxic or potentially toxic).

Since such natural processes occur at a geological time rate, its concentration is at a level of a few parts per million or less, and they are designated as *trace metals*.

Therefore, the increase of the concentration of heavy metals in environmental compartments (atmosphere, lithosphere and hydrosphere) is a consequence of the high demand for industrialized products by population couplet to its high growth rate. In view of that, the consume of industrialized and organic products culminates in the generation of high amount of residues, which are discharged, in most of the cases, without taking into account the sustainability of wastewater treatment processes. Independently of the type of emission, the ultimate fate of metal species is natural waters and soil. Even those associated with particulate matter will be deposited at any time over the surface of water or soil. As the main focus of this chapter is related to procedures applied in the removal of metal species existing in water samples, from now on their behavior will be emphasized taking into account the compartment of hydrosphere (rivers, lakes, oceans, etc.).

In general, metal species in water bodies are divided into two forms: solubilized ions in aqueous phase, or associated with organic matter or inorganic particles. The fact that metal species may be found associated with particles or organic matter does not mean that it is inert and poses no risk to life and environment. Some slight changes in parameters such as pH (hydrogen potential), redox conditions or others physicochemical parameters may be responsible to mobilize adsorbed metal species to water column. Figure 1 illustrates the behavior of particular metal species due to a change in water column pH.

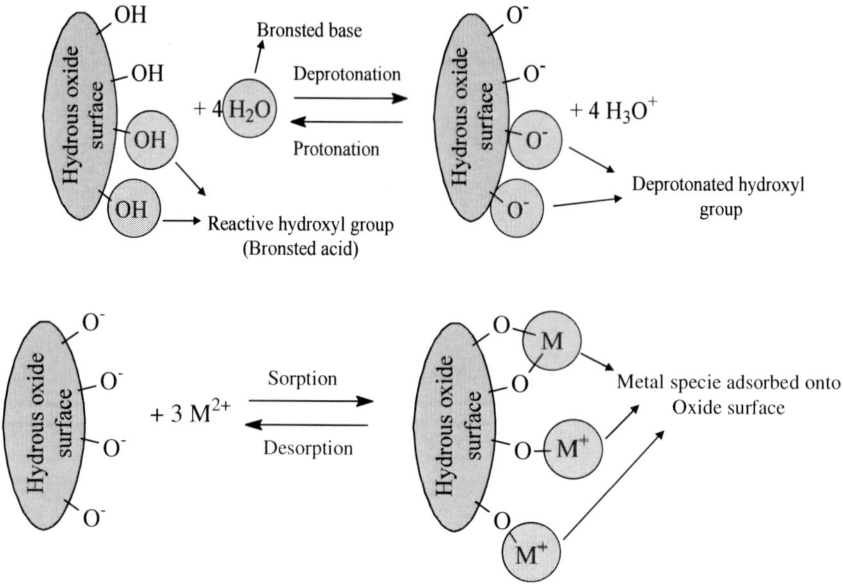

Figure 1. Schematic mechanism of metal adsorption over a sediment particle surface. The influence of pH.

As depicted in Figure 1, hydrous oxide exists in large amounts in natural waters and its reactivity is a consequence of the presence of hydroxyl groups (OH) over the particles as well as its high surface area. The surface area is related to particle size and porosity, the smaller the particles the higher is its surface. In addition, if the surface is high the amount of reactive hydroxyl groups will also be.

Hydroxyl groups play an important role over the particle surface once it behave like Bronsted acids (proton donor) and depending on the environmental conditions it may be deprotonated. When it happens the negatively charged surface may coordinate metal species in solution through a process called adsorption. The term adsorption is defined as the accumulation of matter at the solid-fluid interface (a fluid comprehends liquids and gases) [2], it occurs mainly when physical interactions are involved between adsorbent and the adsorbate (adsorbed species). Other interaction types such as ion exchange and electron pair sharing through covalent bonds are also common.

Natural and Synthetic Adsorbents

Adsorbents or sorbents are materials applied in the adsorption of some substance or element. In chemistry the term *sorbent* and *adsorbent* are applied as synonymous regarding the solid phase, although the adsorption and sorption represents physical and chemical interactions, respectively.

Nowadays we have different types of materials used as adsorbents in the removal of metal species from aqueous samples, whose prices can vary significantly. One of the most used material in water purification is the activated carbon, which can be obtained through controlled burning of some natural products such as biomass residues [3], jute stick [4], chitosan [5], live-mill waste [6] etc., for which the main mechanism involved in the removal of metal is related to pore size exclusion.

From two decades ago until now, other kind of material that has been extensively studied are those based on silica gel. Along with activated carbon, without any doubt, they are the most applied materials in solid-phase extraction. As depicted in Figure 1 for hydrous oxide, silica particles also has its surface covered by hydroxyl groups, and due to its reactivity a series of reactions are possible to improve its adsorption capacity. Along with its reactive surface silica also presents a high surface area and thermal stability [7], which contributed for its extensive use. Differently from activated carbon, it is possible to obtain silica particles with the desired characteristic through many synthetic routes using the sol-gel process. Briefly, this process is characterized by the materials synthesis in which hydrolysis and polycondensation of a reagent (alkoxide) take place yielding a sol (dispersion of colloidal particles in a liquid) followed by a gel [8]. For the past few years, several scientists have been developing new materials through the use of surfactants as pore templates for porous silica [9-13]. These methods employing surfactant are of great interest, once materials with desirable properties (such as high specific surface area, a wide range of pore sizes symmetries and pore volume) can be easily obtained through the adjustment of the synthesis parameters [9-14].

Despite the improvement in material synthesis, metal ion adsorption by pure silica does not occur satisfactorily to turn it into a good adsorbent. This way, in order to convert this material into a special adsorbent, several scientists around the world studied different type of reactions to attach desirable molecules onto its surface. Bellow it is represented some molecules that were used in such reactions.

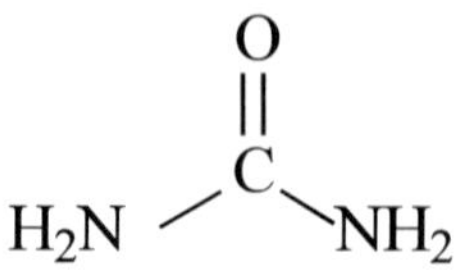

Urea

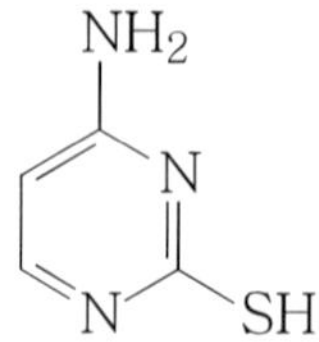

4-Amino-2-mercaptopyrimidine

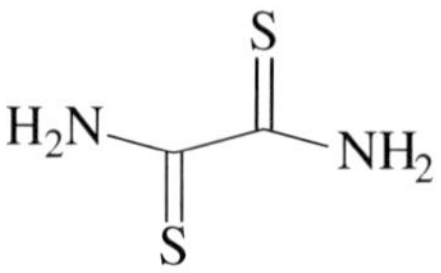

Rubeanic acid

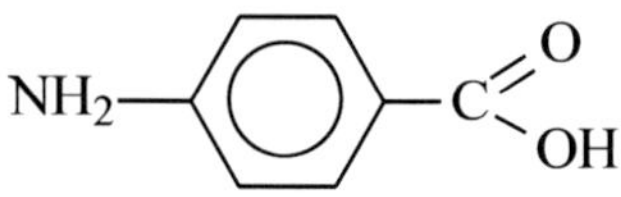

4-Aminobenzoic acid

Figure 2. Molecules containing Lewis-bases applied in silica and cellulose surface modification [15-18].

As can be seen in Figure 2, these molecules present some elements in common, such as nitrogen, sulfur and oxygen, which are classified in chemistry as *Lewis base*. This term tells us that such species are able to share electron pairs with others species in solution. Aiming to clarify how the interaction occurs in a very simple way, we can say that electrons (negatively charged) attract metal species (positively charged) existing in solution and a chemical bond is established.

Although the good adsorption capacities attained by modified materials, all of them exhibit a high production cost once high purity chemicals are required for their syntheses. Furthermore modification reactions are performed under controlled atmosphere and through the use of toxic solvents, such as toluene, xylene, dimethylformamide and others, thus, a new material with environmental purpose is created, though, on the other hand, much toxic residue is generated.

As an alternative to the use of synthetic materials, a great variety of natural materials were also applied in the removal of heavy metals from water samples. The removal mechanism is exactly the same as mentioned for synthetic materials, with the advantage of extremely low costs and similar extraction capacities (sometimes even higher than those presented by synthetic materials). Some examples of natural adsorbents are banana peels [19], coconut coir [20], dead macrophytes [21], waste fruit cortex [22], almond shell [23] and dried cactus [24] among others.

Brief General Considerations for the Application of Biomaterials As Adsorbents

Adsorbents generally are in-powder materials of large superficial area whose surface can interact with molecules, metal ions, and etc. (so-called *sorbates*) in order to retain such species through a superficial interaction, which may be physical (*physisorption*) or chemical (*chemisorption*). Whereas physisorption is merely based on the electrostatic interaction between sorbate and adsorbent, the chemisorption gives rise to a much more stable sorbate-adsorbent interaction, since it is based on the formation of chemical bonds between them, (granting a more effective immobilization of the sorbate by the adsorbent) [25]. It is important to emphasize that, in the case of chemisorption, the existence of chemical affinity between sorbate and adsorbent is fundamental. In view of the interaction between metal-ions and Lewis-bases during complexation, such phenomenon may be exploited to perform the uptake of metal species from aqueous media through the use of an adsorbent material. To do so, it is indispensable that groups of interest (groups that may act as Lewis-bases) be present in the constitution of the adsorbent, so that it can effectively interact with metal ions in aqueous phase. Biomaterials present a great potential to be applied in the uptake of metal ions, once such materials already possesses in their molecular structure groups such as alcohols, amines, amides, sulphydryls, carboxylic etc., which may act as Lewis-bases and perform the complexation of metal ions from aqueous media. Nevertheless, prior to the application of biomaterials as adsorbents, their preparation is a fundamental step to acquire a good adsorbent since its adsorptive properties will be intimately related to some aspects associated to their superficial area and to the accessibility to the material's adsorption sites (which are regulated by the particles morphology) [19, 25-27].

In contrast with many synthesized materials reported in the literature such as modified silica, polymers, cellulose etc. The preparation of biomaterials presents some interesting advantages, such as their preparation through simple processes (namely, drying, pulverizing, grinding and sieving) and the non-production of hazardous chemical residues and solvents (as it often happens for synthetic and modified materials), which is found in agreement with the Green Chemistry's principles. This is due to the fact that biomaterials already present groups of interest in their molecular structure to perform adsorption of metal ions and dispense modification steps prior to their application. The main reason to apply those techniques to biomaterials is to increase their superficial area in order to optimize metal uptake. A disadvantage of biomaterials in this aspect is that hardly ever they will provide specific superficial areas much larger than some few meters square per gram, which is drastically smaller than the areas of silicas, zeolites, activated carbons etc. (which can reach values of hundreds or even thousands of meter square per gram). Despite such drawback, it is not uncommon to encounter biosorbents with high adsorption capacities.

The uptake of metal ions from aqueous media may be exploited even as an analytical tool, as in the decontamination of wastewaters. The first approach makes use of the adsorptive property of a material to retain metals of interest (*analytes*) from an aqueous sample prior to the analysis of metal content. The main advantage of this approach is that it enables the pre-concentration of metals from an aqueous sample (once the amount of analytes from the sample may be retained over the surface of the adsorbent) so that it can be posteriorly eluted with a reduced volume of acidic solution (*eluate*), thus increasing the concentration of the analytes several times in the eluate (detailed explanation on pre-concentration procedures is given in the following topics). The pre-concentration step allows detecting how much metal contaminants there are in some sample of low-concentration by Flame Atomic Absorption Spectrometry (*FAAS*), rather than resorting to more expensive techniques such as Inductively-Coupled Plasma Optical Emission Spectrometry (*ICP-OES*) or Graphite Furnace Atomic Absorption Spectrometry (*GFAAS*). Therefore, it enables the monitoring of the concentration of toxic and potentially toxic metal species such as Cu(II), Ni(II), Cd(II), Pb(II) etc. in natural waters, for instance.

Also, regarding the application of adsorbents in the treatment of contaminated wastewaters, some works report the utilization bed-reactors to perform the decontamination of aqueous residues [28-33]. Such devices consist of reactor vessels containing the adsorbent materials, and in which metals will be extracted from the residues so that they can be safely returned to the environment (or reused) with no risk to human health or wild life.

Preparation of Cassava Husks Powder for the Adsorption of Metal Ions from Aqueous Media

In view of the potentiality biomaterials present as alternative adsorbents for the monitoring and treatment of aqueous residues, cassava husk may be an interesting alternative to be used as an adsorbent material, and to be applied in the uptake of metal species from aqueous media [34]. To do so, it is important to perform the correct preparation of cassava husk, so that it presents maximum efficiency. It is commonly expected from an adsorbent that it presents a high superficial area over which metals will be retained by chemical or physical interactions. Therefore, it is indispensable for cassava husks to be turned into a fine powder, prior to its application in adsorption experiments.

Figure 3. Preparation process applied to cassava husks.

In contrast with most of the synthetic materials, one the main advantage of adsorbent biomaterial lies on their preparation through simple processes, with no need for the use of dangerous chemical products. Several preparation methods are reported for different biomaterials, nevertheless, most of them encompass simple steps such as drying, pulverizing, grinding and/or sieving. The preparation of the cassava husk biosorbent may be found schematically represented in Figure 3. As it is depicted in such figure, cassava roots were washed and peeled to obtain dust-free cassava husks. The husks must be heated in an circulating air oven in a temperature around 80 - 100 °C for at least 24 hour in order to be dried. After drying natural material presents a brittle consistency and may be pulverized in a knife grinder. To obtain a uniform particle size of the material, the powder may be sieved in varied-particle-sizes sieves, and a specific fraction can be selected and applied in adsorption experiments. Usually the smallest fraction is selected due to its high specific surface area. Finally, prior to its application, the obtained material may be washed with acidic solution to remove soluble impregnated organic substances so such compounds could not interact with the adsorbate, and will not compromise the effective metal adsorption over cassava husks or other biomaterial.

Characterization of the Cassava Husks Powder

Characterization of material surface before its application is of great importance. Through this step is possible to infer if material is going to exhibit high of low adsorption capacity. Several techniques are employed to material surface investigation such as Fourier Transform Infrared Spectroscopy (*FTIR*), ^{13}C Nuclear Magnetic Resonance (*^{13}C NMR*) and elemental analysis. The two first mentioned techniques basically gives information about the groups of interest that are present in the material's constitution, whereas elemental analysis provides information about how much H, N, C and S constitutes the material. Also, the material had its superficial area measured through adsorption of nitrogen. In the case of cassava husks, as available in the literature, FTIR analysis indicated the presence of CH, OH and C–O bonds from glycoside units, which were expected since they are the main constituent of cellulose. Also, C=O from carboxylic acids could be verified. Infrared absorption analysis also provided the presence of C–N bonds in cassava husks, which may be assigned to amino acids such as thiamine, methionine etc. In fact, other works reported the presence of methionine, tryptophan and phenylalanine in cassava's constitution [35, 36].

It is important to emphasize that O- and N-based organic groups such as alcohols, carboxylic acids and amines play a crucial role in adsorption processes, since they may act as Lewis-bases by coordinating with metal ions.

^{13}C NMR analysis confirmed CO bonds associated to glycosides units and also pointed out to the presence of apiosyl in cassava husks' structure. No evidence of C≡N bonds indicated that no significant amounts of linamarin and lotaustralin are present in the material. CNH_2 and CN^+ were verified and confirmed the presence of methionine and thiamine. Moreover, cysteine could not be detected as a constituent of cassava husks [34]. According to literature, cassava husks presents considerable amount of nitrogen, as can be observed in Table 1 [34].

Table 1. Composition of C, N, S and H in cassava husks

Element	Composition (%)
C	41.42
H	6.64
N	1.02
S	0.60

This characterization is of great importance once the obtained results may be used to attest the compounds that have been suggested through other characterization techniques. This characterization is so important that it should be the first to be performed, since the absence of elements containing free electron pairs to be shared (Lewis base) is strong evidence that the material will not perform well as metal ion adsorbent.

Point of Zero Charge (PZC)

Only with full knowledge of the material surface is possible to obtain some success and correctly interpret the results obtained in an adsorption process. The determination of material PZC is of great importance as the results will be used to predict at which pH value the material will exhibit better adsorption capacity and at which pH the surface of the material becomes electrically neutral. Therefore, at pH values different from pH_{pzc}, the material can be either positively or negatively charged, and the surface charge will be determined by organic groups protonation and deprotonation phenomena.

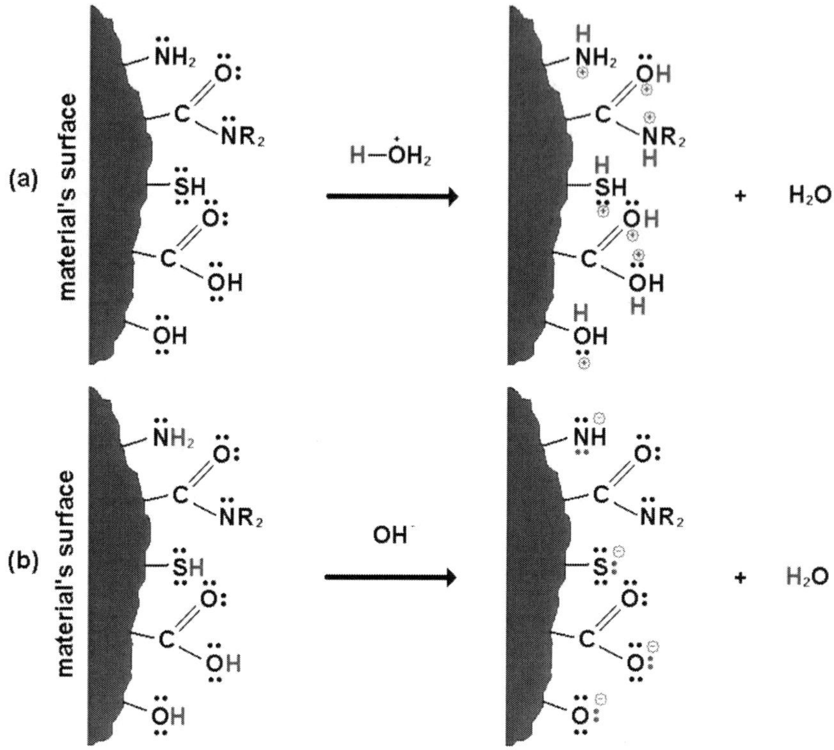

Figure 4. Protonation (a) and deprotonation (b) of organic groups, and the formation of electric charges over cassava husks' surface.

Figure 4 illustrates such phenomena for a cassava husks particle. In Figure 4 (a), we see the protonation of cassava husks, which occurs at low pH values. Under such conditions, the high hydronium concentration in the medium ($[H\text{-}OH_2]^+$ over the arrow) will donate H^+ species to the non-bonding electrons of organic groups such as amines, amides, sulphydryls, carboxylic, alcohols etc., thus, such species besides occupying the adsorption sites of metal cations will also attribute positive charges to the material surface. This phenomenon will cause the repulsion of metal ions and will not allow them to be effectively adsorbed over the material's surface. On the other hand, too elevated pH will imply in a high OH^- concentration. The OH^- species in the medium may extract H^+ from organic groups, and negative charges will be formed over the material's surface, as demonstrated in Figure 4 (b). Despite one may suppose that negative charges over cassava particles may act in favor of the adsorption of cations due to the attraction of cations by the negative charges, OH^- species in the matrix may involve metal cations (*solvatation*) and prevent them to be adsorbed as well. Also, under too elevated OH^- concentration, the metal ions may even be precipitated as solid crystals, not allowing them to be adsorbed at all. Therefore, the ideal pH range to perform adsorption of free-cations is the one found closest to the material's pH_{pzc}. For cassava, pH_{pzc} obtained was 5.2. At pH values lower than this value, cassava husks' surface will be found positively charged, whereas for higher pH values, it will be negatively charged.

Moreover, since cassava's pH_{pzc} found is acidic, it implies that acidic organic groups are predominant over the material's surface with relation to basic groups.

Batch Adsorption Experiments

The batch experiments consist of shaking a certain volume of metal solutions with a determined mass of the adsorbent material into stopped vials. The agitation gives rise to a suspension of adsorbent's particles, what enables an effective contact between the particles and metal solution in order to extract metal cations from the matrix. Such experiments provide information about metal uptake for a determined mass of material under controlled medium conditions.

An example of agitation system that may be used to perform the batch experiments with cassava husks or other adsorbents is depicted in Figure 5. The first batch experiment that may be performed is to determine the kinetic equilibrium, where a fixed concentration and volume of metal solution is stirred with a fixed mass of adsorbent for a time interval, for instance, from 1 to 250 minutes. According to the literature cassava husks presents a very rapid kinetics and equilibrium between metal ions in solution and adsorption sites was attained within 5 minutes for copper. The rapid equilibrium kinetic is an advantage for adsorbent materials, once they can be successfully applied in preconcentration experiments, where the material is packed in a small column and the solution is pumped through it.

Another parameter of great importance is the metal solution pH. In this batch experiment the solution volume and metal concentration is kept constant as well as the adsorbent mass. The main objective here is to determine at which pH the adsorption is favored. The results obtained may be correlated with those obtained from the PZC experiment. An example of batch experiment as a function of pH is depicted in Figure 6 (a). As can be observed from Figure 6 at pH 1, the adsorption of Cu(II) ions was practically null, but the adsorption increased to reach a plateau in the pH range from 3 to 6. This behavior is very typical for many adsorbents, and may be explained by protonation, and deprotonation effects, as illustrated by Figure 4, and discussed previously. At lower pH, H^+ ions compete with metal species in the medium and, due mainly to their high concentration, are immobilized to cassava husks surface, causing the repulsion of Cu(II) cations. As pH increases, this effect is reduced and the metal ions can be adsorbed by cassava husks more effectively. In addition, the optimum pH range for the adsorption of Cu(II) ions indicates that cassava husks may be directly employed to extract such species from natural water samples as well as from certain effluents.

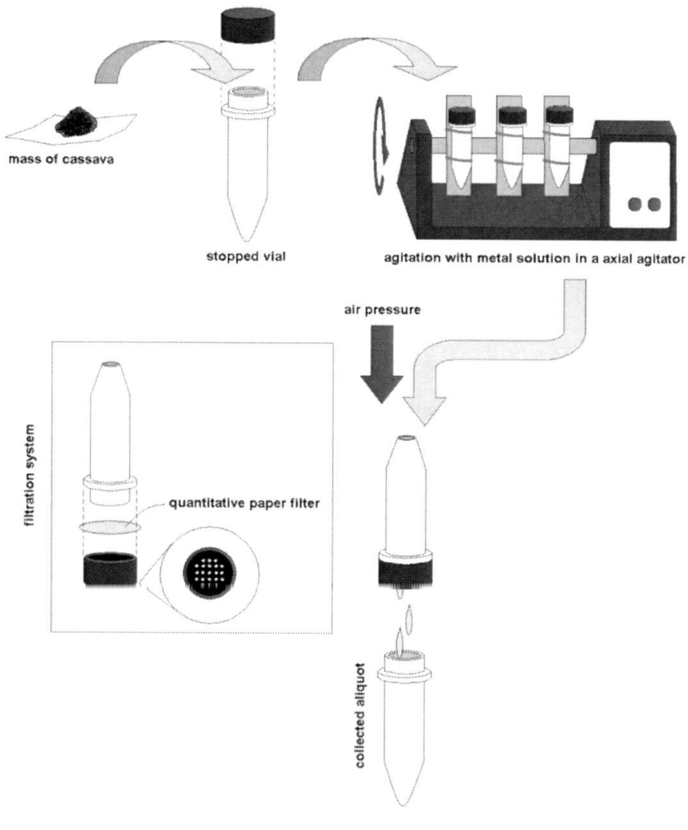

Figure 5. Batch agitation system.

Table 2. Comparison between the maximum adsorption capacities of different materials

Material	Maximum adsorption capacity (mmol g^{-1})	Reference
Cel-DTX	0.058	[37]
Peanut hull pellets	0.18	[38]
AMP modified silica	0.45	[16]
GMZ bentonite	0.10	[39]
p-Aminobenzoic modified cellulose	1.96	[40]
Amidoamidoxime silica	0.016	[41]
Cassava husks powder	0.137	[34]

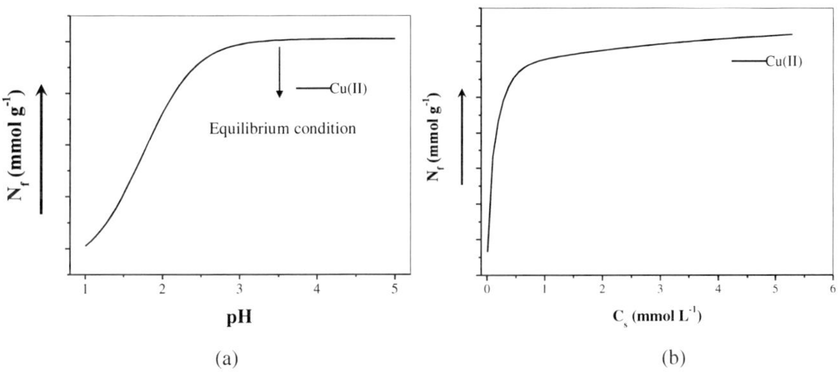

Figure 6. Adsorption of Cu(II) as a function of pH (a) and Cu(II) concentration (b).

The experiment performed to investigate the maximum adsorption capacity resulted in a graph as the one represented by Figure 6 (b). From such figure we see that as Cu(II) concentration increased, metal uptake also increased till the moment in which a plateau is formed. When the adsorption of Cu(II) tends to a constant value, its maximum adsorption capacity is attained, and practically all cassava husks' surface is covered by Cu(II) ions. The maximum adsorption capacity found for cassava husks was of 0.137 mmol g^{-1}. Aiming to compare the material's performance with other studied adsorbents, Table 2 was built.

Pre-concentration Experiments

Importance for the Environment and Human Health

There are some chemical compounds, mainly metal ions, which may develop double effect over organisms' physiology. They may even be essential if their concentration is found at an optimum concentration range, or toxic, if their concentration exceeds such levels. As an example, Co, Cr, Cu, Fe, Mn and Se participate in some metabolic pathways and also are important constituents of some proteins and vitamins, such as Fe-hemoglobin and vitamin B-12 [42]. On the other hand, other types of metal species are considered toxic even at very low concentrations, as in the case of As, Cd, Pb and Hg, since they do not present any biological known function for any organism. Individuals poisoned by some of these species may suffer from serious health problems, such as nervous system disorders, diverse types of cancer or even death [43].

Metals present in the environment may be dispersed from several anthropological or natural sources. Industries are the greatest responsible for anthropological metal emissions, thus contaminating air, soil or hydric resources. Metals were always employed in large scale in the production of a great variety of products, and, therefore, they are present in most of the residues produced in the manufacturing processes. During several decades pollution regulations established by environment monitoring entities was insufficient or none. Only in the recent years pollution emissions have triggered a deeper discussion about environmental preservation, mainly due to the scientific advance in such field. As time passed by conscientization about wastes treatment became vital, and several treatment methods were developed and applied in remediation of industrial residues. Nevertheless treatments for metal decontamination may be highly complex since some metals present elevated reactivity and are not easily degraded.

Risks posed to human health by metal contaminants are extensive and are mainly related to the concentration and bioavailability of such species in the environment. Table 3 summarizes some of the damages caused by metal contaminants and the industrial processes associated to their generation.

Table 3. Metal sources and their effects over human health

Metal	Source (industrial process)	Effect to human health
Cadmium	Mining, PVC production, dyeing, metal alloys production, batteries production, also present in phosphatized fertilizers [44].	Arterial hypertension, kidney damage, destruction of testicular tissue, cancer among others [44-46].
Lead	Fossil fuel burning, iron and steel production[5]. Lead is utilized in the production of lead acid batteries, welding, metal alloys, cable coating, dyes, rust inhibitors, ammunition, plastic stabilizers [46].	Inquietude, irritability, lack of attention, headaches, trembling, abdominal colic, kidney and nervous system damages, hallucinations, loss of memory and encephalopathy [44, 46].
Mercury	Fossil fuel burning, gold mining, cement production, lamp production [5]. Batteries production, thermometers, dental obturation, and, previously, in fungicides [47].	Serious respiratory irritation, pneumonite, pulmonary edema, damages to the brain, nerves, kidney, lungs and, in some extreme cases, it may cause coma and death [46]. Also, neurological damages, including irritability, paralyzation, blindness and madness [47].
Arsenic	Fossil fuels combustion, in particular mineral charcoal, contamination of some early phosphorous-based pesticides, mining wastes as byproducts from the extraction of copper, gold and lead [47].	Abdominal pain, vomit, diarrhea, muscular pain and weakness, eruption, sensitivity loss in the peripheric nervous system, cancer in the skin, bladder and lungs [44, 46].

In view of the risks posed by metal contaminants, nowadays, most of the governmental entities for environmental protection began to take actions to avoid so many environmental and health damages. The major of them was the development of regulations that established rigorous contamination limits for the most varied natural resources. Such measurement aimed to minimize impacts over the environment caused by mankind in order to protect public health, ecological balance and the well-being of society. With that in mind, the development of simple, rapid and efficient analytical techniques and methodologies for the monitoring of the here-mentioned substances has acquired major importance, once the demand for their qualitative and quantitative control in environmental, agronomic, biological and food samples has increased significantly.

The rigid contamination limits imposed by regulation entities and the recent technological advances lead to the development of highly sensitive analytical techniques with very low detection/quantification limits [for instance, Inductively-Coupled Plasma Optical Emission Spectrometry (ICP-OES) and Graphite Furnace Atomic Absorption Spectrometry (GFAAS)] for the determination and monitoring of metal contamination in water samples. Nevertheless most of such techniques frequently are associated with elevated equipment, installation and operational costs as well as an extensive specialization of the analysts to handle different samples and interpret results. Consequently few laboratories may present adequate infrastructure to perform trustworthy analytical results, and under-development countries usually end up suffering from the lack of residue/water monitoring. Despite of the inconvenience associated to ICP-OES and GFAAS techniques, Flame Atomic Absorption Spectrometry (FAAS), which is a cheaper, simpler and more disseminated technique for determination of metal cations in solutions, may still be used in the determination of metal contents in aqueous samples as long as pre-concentration procedures be applied to the samples prior to the analyses. Thus, the lower efficiency of FAAS technique (related to its higher detection/quantification limit in comparison to ICP-OES and GFAAS) to analyze trace and ultra-trace elements in water samples or residues can be circumvented and become an economically viable alternative to detect and quantify toxic and potentially toxic metals.

Preconcentration Methodology

One of the main pre-concentration methods applied for analytical purposes is based on the use of a column packed with an adsorbent, which will retain the analyte of interest (in this chapter, it will be discussed specifically the in-flow pre-concentration of metal species). A scheme to represent a pre-concentration system may be found in Figure 7.

Pump 1 in the system scheme will be responsible to percolate a determined amount of metal solution through the column containing the adsorbent material. As the solution passes through the column, the metal ions get trapped inside it, and can only be remove by a flow of acidic solution (*eluent*), which is provided by *Pump 2*. A flow commutator is generally employed to perform the flow shift (demonstrated in Figure 7 by the shift in the commutator lever from position 1 to 2), and a reduced volume of eluent is collect to be analyzed through FAAS (alternatively it may be directly introduced to the equipment). It is important to emphasize that the pre-concentration efficiency depends heavily on the volumes of both metal solution and eluent (but not restricted only to these parameters). With that in mind, it is safe to claim that the greater the volume of the sample in comparison to the volume of the eluent, the greater the pre-concentration will be, since the amount of metal contained in the sample will be transferred to a reduced volume of eluent. Thus, a metal content from a sample may be magnified from tens- to hundreds-fold by this method, though, many others parameters must be taken into consideration in order to make pre-concentration viable. A previous study about experimental conditions should be made, such as optimum flow rates, eluent concentration, sample and eluent adequate volumes, adsorbent mass etc.

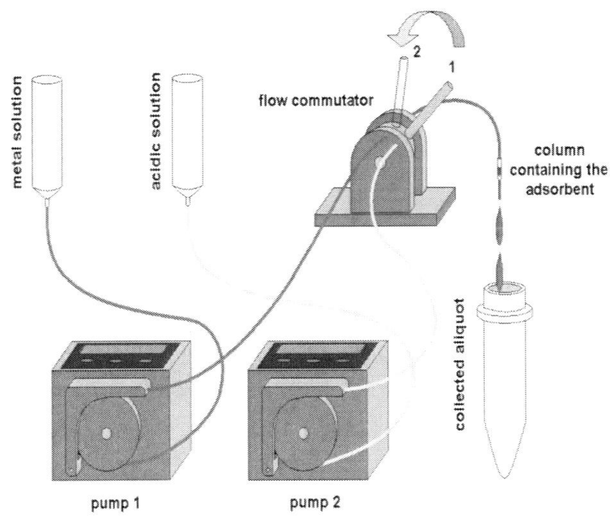

Figure 7. Pre-concentration system.

Regarding the pre-concentration experiments utilizing cassava husks, 50 mL of sample solutions (whose pH was previously adjusted to the range 5 – 6) were percolated through column containing the adsorbent material. The retained Cu(II) ions were later removed from the column with 1 mL aliquots HNO_3 solution, which were analyzed via FAAS.

Jorgetto et al. described in their paper [35] the application of cassava husks in the pre-concentration of Cu(II) from a natural water sample and a tap water sample. Cu(II) concentrations for the respective samples were found to be 1.808 ± 0.039 and 36.87 ± 0.13 μg L^{-1}. In order to validate the methodology, the procedure was also applied to a certified reference water sample and the concentration obtained was 23.03 ± 0.79 μg L^{-1}, whereas the certified reference concentration was 22.7 ± 0.31 μg L^{-1}. This demonstrated the real effectiveness of pre-concentration methodology for the determination of trace metal in water samples. In addition, another outstanding property verified for cassava husks was its high chemical stability, which could be verified through cassava husks high endurance towards several cycles of adsorption and desorption (column reutilization experiment). Also related to this aspect, since the material is undergone to several cycles of adsorption and desorption, the elution step of each cycle with acidic solution may degrade the adsorbent, compromising its efficiency, nevertheless, cassava husks kept its performance during the whole range of reutilization experiment studied (40 adsorption-desorption cycles), indicating that its reutilization capacity probably is greater than this number of cycles.

Final Considerations

Each material has its own characteristics, which are directly associated with their chemical composition. Cassava was studied in detail and showed exceptional results regarding metal removal in water samples. As observed for other materials, it can be applied for environmental purposes with an advantage that only the husks, which are usually discharge, are used as adsorbent material. The effective cost to produce cassava husks powder to be applied as filter bead is extremely low and it can be obtained practically anywhere in world. The objective of this work was not to solve the world's problem regarding environmental contamination by metal species, it was to give the reader another point of view about the possible use of organic materials. Cassava husks is only an example, there are infinity of other products that could be used for the same purpose. Despite of this, its use as household filter should be avoided as several chemical experiments and characterization procedures are required to attest whether such application is viable and safe.

REFERENCES

[1] S. L. Stipp, *Molec. Sim.* 28, 497 (2002).
[2] W. Stumm, J. J. Morgan, *Aquatic Chemistry*: Chemical equilibria and rates in natural waters, Wiley Interscience, New York (1996).
[3] M. Gueye, Y. Richardson, F. T. Kafack, J. Blin, *Journal of Environmental Chemical Engineering* 2, 281 (2014).
[4] M. Asadullah, I. Jahan, M. B. Ahmed, P. Adawiyah, N. H. Malek, M. S. Rahman, *J. Ind. Eng. Chem.* 20, 896 (2014).
[5] S. Hydari, H. Sharififard, M. Nabavinia, M. R. Parvizi, *Chem. Eng. J.* 194, 282 (2012).
[6] M. I. Bautista-Toledo, J. R. Rivera-Utrilla, R. Ocampo-Pérez, F. Carrasco-Marín, M. Sánchez-Polo, *Carbon* 73, 350 (2014).
[7] R. K. Iler, The chemistry of silica, John Wiley & Sons, New York (1979).
[8] L. L. Hench, J. K. West, *Chem. Rev.* 90, 72 (1990).
[9] M. A. U. Martines, D. R. Carmo, G. R. Castro, L. Caetano, *J. Sol-Gel Sci. Technol.* 59, 193 (2011).
[10] J. Aguado, J. M. Arsuaga, A. Arencibia, M. Lindo, V. Gascón, *J. Hazard. Mater.* 163, 221 (2009).
[11] Z. A. Alothman, A. W. Apblett, *J. Hazard. Mater.* 182, 590 (2010).
[12] S. Hao, Y. Zhong, F. Pepe, W. Zhu, *Chem. Eng. J.* 189, 167 (2012).
[13] A. Shahbazi, H. Younesi, A. Badiei, *Chem. Eng. J.* 168, 518 (2011).
[14] S. M. Evangelista, E. Oliveira, G. R. Castro, L. F. Zara, A. G. S. Prado, *Surf. Sci.* 601, 2202 (2007).
[15] M. Benitez, D. Das, R. Ferreira, U. Pischel, H. García, *Chem. Mater.* 18, 5603 (2006).
[16] A. S. Pereira, G. Ferreira, L. Caetano, M. A. U. Martines, P. M. Padilha, A. Santos, G. R. Castro, *J. Hazard. Mater.* 175, 403 (2010).
[17] O. Jorgetto, R. I. V. Silva, M. M. Longo, M. J. Saeki, P. M. Padilha, M. A. U. Martines, B. P. Rocha, G. R. Castro, *Appl. Surf. Sci.* 264, 374 (2013).
[18] G. R. Castro, J. D. Oliveira, I. L. Alcântara, P. S. Roldan, C. C. F. Padilha, A. G. S. Prado, P. M. Padilha, *Separ. Sci. Technol.* 42, 1340 (2007).
[19] R. S. D. Castro, L. Caetano, G. Ferreira, P. M. Padilha, M. J. Saeki, L. F. Zara, M. A. U. Martines, G. R. Castro, *Ind. Eng. Chem. Res.* 50, 3451 (2011).

[20] M. H. Gonzalez, G. C. Araujo, C. B. Pelizaro, E. A. Menezes, S. G. Lemos, G. B. Souza, A. R. A. Nogueira, *J. Hazard. Mater.* 159, 256 (2008).

[21] P. Miretzky, A. Saralegui, A. F. Cirelli, *Chemosphere*, 62, 254 (2006).

[22] K. Keely-Vargas, M. Cerro-Lopez, S. Reyna-Tellez, E. R. Bandala, J. L. Sanchez-Salas, *Phys. Chem. Earth*, 39, 29 (2012).

[23] A. Ronda, M. A. Martín-Lara, E. Dionísio, G. Blázquez, M. Calero, *J. Taiwan Inst. Chem. E.* 44, 473 (2013).

[24] N. Barka, M. Abdennouri, M. El Makhfouk, S. Qourzal, *J. Environ. Chem. Eng.* 1, 149 (2013).

[25] M. Chikazawaand, Third Edition. Edited by Hiroaki Masuda, Ko Higashitani, and Hideto Yoshida. CRC Press 2006.

[26] E. Marañón, H. Sastre, *Bioresour. Technol.* 40, 73 (1992).

[27] R.G. Pearson, *J. Am. Chem. Soc.* 85, 3533 (1963).

[28] H. Muhamad, A. H. Doan, A. A. Lohi. *Elsevier. C Oxford.* 158, 369 (2010).

[29] K. Naddafi, R. Nabizadeh, R. Saeedi. *J. Hazard. Mater.* 147, 785 (2007).

[30] K. H Chu, *J. Chem. Eng.* 97, 233 (2004).

[31] A. Chatterjee, S. Schiewer, *J. Chem. Eng.* 244, 105 (2014).

[32] M. E. R. Carmona, M. A. P. Silva, S. G. F. Leite, O. H. V. Echeverri, C. O. Campo-López, *J. Taiwan Inst. Chem. Eng.* 43, 428 (2012).

[33] M. Calero, F. Hernáinz, G. Blázquez, G. Tenorio, M. A. Martín-Lara, *J. Hazard. Mat.* 171, 886 (2009).

[34] A. O. Jorgetto, R. I. V. Silva, M. J. Saeki, R. C. Barbosa, M. A. U. Martines, S. M. A. Jorge, A. C. P. Silva, J. F. Schneider, G. R. Castro, *Appl. Surf. Sci.* 288, 356 (2014).

[35] N. L. R. King, J. H. Bradbury, *J. Sci. Food Agric.* 68, 223 (1995).

[36] Nambisan, *Food Chem. Toxicol.* 49, 690 (2011).

[37] A. O. Jorgetto, R. I. V. Silva, M. M. Longo, M. J. Saeki, P. M. Padilha, M. A. U. Martines, B. P. Rocha, G. R. Castro, *Appl. Surf. Sci.* 264, 368 (2013).

[38] P. D. Johnson, M. A. Watson, J. Brown, I. A. Jefcoat. *Waste Manage.* 22, 471 (2002).

[39] J. Li, J. Hu, G. Sheng, G. Zhao, Q. Huang. *Colloids Surf. A.* 349, 195 (2009).

[40] G. R. Castro, I. L. Alcântara, P. S. Roldan, D. F. Bozano, P. M. Padilha, A. O. Florentino, J. C. Rocha. *Mater. Res.* 7, 329 (2004).

[41] W. Ngeontae, W. Aeungmaitrepirom, T. Tuntulani, A. Imyim. *Talanta*, 78, 1004 (2009).

[42] H. G. Seiler, H. Sigel, New York, Marcel Dekker, Inc, p.79, 1988.
[43] L. Tsalev, Z. K. Zaprianov, Flórida: CRC Press, p.87, 1983.
[44] L. Järup. *Brit. Med. Bul.* 68, 167 (2003).
[45] R. Angeletti, G. Binato, M. Guidotti, S. Morelli, A. A. Pastorelli, E. Sagratella, S. Ciardullo, P. Stacchini, *Chemosphere,* 100, 83 (2014).
[46] WHO - Guidelines for Drinking-water Quality, World Health Organisation, Fourth Edition, (2011).
[47] S. E. Manahan, Boca Raton: Lewis Publishers, 1997.

In: Cassava
Editor: Francis P. Molinari

ISBN: 978-1-63321-031-8

Chapter 2

FUNCTIONAL AND NUTRITIONAL CHARACTERIZATION OF CASSAVA FLOURS FOR INDUSTRIAL APPLICATIONS

L. J. Sívoli[1,*], A. T. Ciarfella[2,†] and E. E. Pérez[3,‡]

[1]Departamento de Ciencias Biomédicas, Facultad de Ciencias Veterinarias, Universidad Central de Venezuela

[2]Unidad de Cursos Básicos, Núcleo de Anzoátegui, Universidad de Oriente

[3]Instituto de Ciencia y Tecnología de Alimentos, Facultad de Ciencias Universidad Central de Venezuela

ABSTRACT

Since, cassava root in its composition have not proteins forming gluten, the functional properties of cassava flour are not sufficient for elaboration of dough for baked goods, bread and pasta. However, if this raw flour is thermal and enzymatically modified, its properties could be adequate in order to develop different products. The goals of the study were to evaluate the nutritional and functional profiles of cassava flours that were elaborated from the edible portion of five clones of cassava a germplasm bank: Three sweets (≤50ppm cyanide) and two bitters (>50 ppm cyanide), and modify one of this flour by using thermal and

* E-mail: sivoli.lilliam@gmail.com.

† E-mail: anaciarfellaudo@gmail.com.

‡ E-mails: perezee@hotmail.com, elevina.perez@ciens.ucv.ve.

enzymatic treatments, with and without addition of stearoyl-2-lactylate (SSL). Cassava flour produced by dehydration (60 °C x 24 hours) was modified: by boiling and autoclaving, in aqueous solutions of flours of 10% (with 2% NaCl), and 33 %, respectively, and dextrinizated with Termamyl®. The results showed differences in crude protein and ash contents in the different varieties of cassava flours assayed. They are also showed that the yield was less for the bitter variety than in the sweet cassava variety. The contents of moisture, crude protein, crude fat, ash, and dietary fiber was not function of the type of roots, varying from 8.1 to 10.6%; 1.3 to 3.5%, 0.4 to 1.1%, 2.0 to 2.6%, and 2.3 to 6.1%, for sweet and bitter varieties, respectively. Carbohydrate content was similar in all flour, exceeding the 80% in all of them. The color profile has also varies with cassava variety. Data are showing that the use of SSL, NaCl and the dextrinization treatment affect the moisture and ash contents significantly ($p \leq 0.05$). The use of SSL decreased the dehydration effect on the flours during the drying; the NaCl use and dextrinization treatment increased the ash content. The flours modified thermically and enzimatically were darker than the raw flour. The particle sizes were also affected by the treatment and SSL addition, showing a granular size varying from 125 to over 250 microns. The combination of dextrinization and SSL increased the granular size beyond 125 microns. Except for the dextrinizated flours (with and without SSL), which have not developed an amylographic profile, the addition of SSL have also decreased the overall viscosity, measured by using the amylograph. The boiling and retorting treatment have produced a type IV farinogram-curve. The SSL addition changed drastically this curves to those of type I, and the enzymatic treatment to the type II. These flours could be used for baked goods manufacture and could also be blended with type IV flour to the bread making. The NaCl addition showed poor effects in the farinographic profile. The modified flours showed total granular disaggregation when were analyzed by SEM, indicative of total gelatinization. As conclusion, it suggest the feasibility to use these flours for the production of numerous baked goods, bread, and pastas gluten-free, since, they showed a wide spectrum of nutritional and functional properties by effect of additive and the treatments applied.

Keywords: Cassava flour, farinographic profile, gelatinization, free-gluten flour

1. INTRODUCTION

Roots of the cassava plants (*Manihot esculenta* C.), beside to be the main staple food in South America and Africa, are in transition from a starchy staple food to an ingredient; such as flour, for uses in food and feed. Traditionally, cassava flour can be produced from washed or peeled roots, which are grated, chipped, or sliced, then sun-dried on trays, and finally milled into flour (Aryee et al., 2006; Pérez et al., 2007, Westby, 2002). Grated roots may also be pressed to remove water excess, and then toasted to produce roasted cassava flour, with up to 95% of gelatinized starch (Tivana, et al., 2010). The toasted flour, that is called "farinha" in Brasil, and is used to produce "farofa" when mixed with egg, meat, vegetables. The grated roots from bitter cassava is used to produce "casabe" or cassava bread by a pre-columbian recipe of the indigenous people (Ciarfella et al., 2013), and to produce "gari", when is roasted in a pan, in some Africa countries, once is it submitted to traditional methods, that involve fermentation, to decrease the levels of cyanide. The fermentation stage improves the internal stability of the starch granules, reduces the swelling power, and decreases the amylose release during heat treatment (Numfor et al., 1998). Additionally the flour can be pregelatinized by using thermal treatment and produce modified flours (Pérez et al., 2007) for different uses. Subsequently, cassava flours is one of the primary products of raw cassava root that has continued to find wider food application (Wang et al., 2012 ; Cueto, et al., 2012, Charles et al., 2007; Banea et al., 2013; Charoenkul et al., 2011; Jisha,et al., 2008; Shittu et al., 2007a, Shittu et al., 2007b; Oboh et al., 2005).

Due, the lack of gluten in cassava flour its functional properties to form viscoelastic dough that retains gas during the leavening are not sufficient for elaboration of dough for baked goods, bread and pasta. The low protein content (Polyorach et al., 2013), lack of gluten (Pasqualone et al., 2010) and its cyanide content (Santana, et al., 2002) in some varieties of cassava are disadvantageous for its use in new product development. The lack of gluten can be overcome through the modification of the raw flours or using additives (Osman, 1961). The low protein concentration has been solved by the incorporation of some protein concentrates, or flours from legume, and fish, being it's mainly use as composite flour (Ciarfella et al., 2013, Fiorda et al., 2013; Nindjin et al. 2011; Jisha et al., 2008; Khalil et al., 2000; Casier et al., 1979; Kwee et al., 1969). Finally, the high cyanide content in some varieties can be overcome by selecting, by analytical procedure, of those varieties considered as innocuous in regards of its cyanide concentration. The main studies on bread quality elaborated from composite cassava-wheat flour, have investigated only the influence of the baking process, the cassava genotype, and effect of an added hydrocolloid, xanthan gum or malting (Ouraga-Djoussou and Bukanga, 1998, Khalil et al., 2000, Westby 2002, Shittu et, al.

2007a,b, Shittu et, al. 2008, Shittu et, al. 2009, Nindjin et al., 2011). Very little attention has been paid to investigate the effect of differently processed cassava flour as ingredients for partial or total substitution in composite flour for food-making. Though, researching in this area is dealing continuously in order to reach its improvement, with easily and adequate technologies (Nambisan, 2011; Bradbury and Denton, 2010; Cumbana et al., 2007; Cardoso et al., 2005; Muzanila et al., 2000; Mark et al., 1989). Moreover, nature has developed cassava varieties with minimal cyanide content, which can be eaten without toxicity risk (FAO, 1999). Using emulsifiers is other assayed option by the baked goods industry for improves the quality of the products prepared by using composite or non conventional flours (Osman et al., 1961, Osman, 1975; Chung and Tsen 1975, 1977). The characteristics which the processor is expecting from an emulsifiers are: improved dough handling including greater dough strength; improved rate of hydration and water absorption greater tolerance to resting time, shock and fermentation; improved crumb structure: finer and closer grain, brighter crumb, increased uniformity in cell size; improved slicing characteristics of bread; crust thickness; emulsification of fats and reduction of shortening; improved symmetry; improved gas retention resulting in lower yeast requirements, better oven spring, faster rate of proof and increased loaf volume; longer shelf life of bread (Kohajdová et al., 2009). Some emulsifiers, i.e., sodium stearoyl-2 lactylate (SSL) show dual properties for both dough strengthening and crumb softening (Stampfli and Nersten, 1995). The use of NaCl is another option to improve the functional properties of the dough from non conventional flours. It have been postulated that the ion presence from sodium salt act on the viscosity and the temperature of maximum viscosity is reached (Osman, 1975, Wootton and Bamunuarachchi, 1980).

The increasing interest in the use of cassava roots for food and industrial purposes especially in the baked goods industry is leaded with total or partial substitution of wheat flour with cassava flour and/or other kind of flours. However, it will be necessary before, the evaluation of the effects of the any applying treatment in the elaborated flour, on the behaviors of the starch (amylograph curves) and protein (farinograph curves), or even on its proximate composition and microstructure. On the other hand, there are consumers that cannot eat product containing gluten. It is of consensus that products from wheat are the representative of the baked goods, breads, and pastas, so these consumers cannot enjoy these foods. Cassava flour is gluten-free; their rheological and granular properties are not satisfactory for elaboration of baked goods, breads and pastas. These are the rheological, physical and colloidal properties developed by the kneading processes, when flour with gluten is used (Cueto et al., 2012). Then, it is postulated a hypothesis that thermal or enzymatic treatments could modify the cassava flour properties in order to make it adequate for elaboration of baked goods, breads and pastas, and its properties could be predicted by analyzing the amylographic and farinographic curves. In that case, the aims of the study were to evaluate the rheological profile, composition, physical properties, and ultrastructure of cassava flour modified by using different heat and enzymatic treatments; all of them, with and without addition of stearoyl-2-lactylate (SSL), and to propose as ingredient for gluten-free food development.

2. Material and Methods

2.1. Materials

Roots of three sweet varieties of cassava (*Manihot esculenta* Crantz, cvs: Tempranita (commercial), UCV-2306, and UCV-2593), and two bitter varieties (UCV-2192, UCV-2681) with commercial, economic and culinary potential of the national collection from the germplasm bank of the Faculty of Agronomy, Universidad Central de Venezuela, were obtained.

Sodium stearoyl-2-lactylate (SSL) Brand: EMPLEX (FFC, 1981).

Alpha-amylase; Termamyl® 120 L, Type L. Declared activity 120 KNU-T/g. pH 6.5, 85-95°C.

2.2. Methods

2.2.1. Hydrogen Cyanide in Roots

For the quantitative determination of cyanide in roots was used the modify method of Essers (1993) using isothiocyanate, and 1,3-dimetiltiobarbiturato as color reagent, instead of the original Cooke (1978) prepared from pyridine and pyrazolone. Cyanide extraction was performed following the Cooke (1978). On the extract, obtained total and free cyanide were determined. Absorbance was measured at 605 nm and for calculates the total cyanide concentration a standard curve of from 0.5 to 2.0 μgHCN/mL was prepared. The linamarase enzyme (EC 3.2.1.21) for the determination of total cyanide was obtained from the root cortical parenchyma (Cooke, 1978).

2.2.2. Preparation of Raw Cassava Flour

Raw cassava flour was prepared following the procedure described by Pérez et al., 2007 and its yield was measured by the follow equation:

$$\%\,Yield = \frac{\text{weight of dry cassava edible pulp}}{\text{weight of fresh cassava edible pulp}}$$

2.2.3. Preparation of Gelatinized Cassava Flours

Raw cassava flour was used for preparing four different gelatinized cassava flours (Figure 1), as follows.

2.2.4. Control Flour with and without SSL

Control flour *with and without SSL* (flour C1 and C2, respectively) were prepared by mixing one part of raw flour and 1 part of water (1:1 flour:water, w/w) and 0.5% of SSL (for C1) in a homogenizer (Cowles Dissolver, Model 1 V-G, Serial LA 266, The Cowles Dissolver Co., Inc., Cayuga, New York, USA). The mixture was dried in a drum drier (Sterling Speedtrol, Model 20, serial 810, Sterling Power System, Inc., Indiana, USA) at 152°C (60psi) during 9sec, and milled in the hammer mill (Fitz Mill Comminuting Machine, Model D, The Fitzpatrick Company; Inc., Chicago, USA) with a 00 sieve (0.5mm).

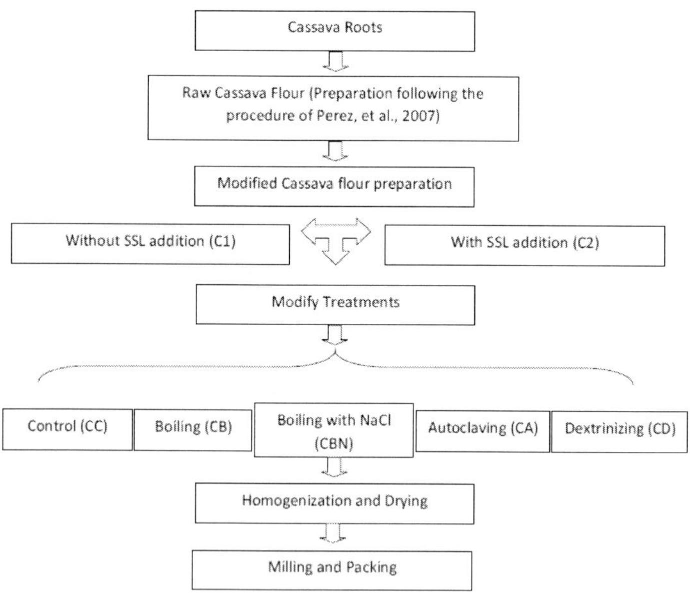

Figure 1. Procedure for the elaboration of control and modified Cassava flours.

2.2.5. Gelatinized Flour by Boiling at 98°C in Aqueous Water System with and without 0.05% of SSL

Gelatinized flour in a water system (flours CB1, with SSL, and CB2, without SSL) were prepared by mixing one part of raw flour and ten part of water (1:10 flour: water, w/w) and 0.5% of SSL (only for CB1) in a homogenizer (Cowles Dissolver, Model 1 V-G, Serial LA 266, The Cowles Dissolver Co., Inc., Cayuga, New York, USA), and gelatinized (98°C) for 30min.

The mixture was dried in a drum drier (Sterling Speedtrol, Model 20, serial 810, Sterling Power System, Inc., Indiana, USA) at 152°C (60psi) during 9sec, and milled in the hammer mill (Fitz Mill Comminuting Machine, Model D, The Fitzpatrick Company; Inc., Chicago, USA) with a 00 sieve (0.5mm).

2.2.6. Gelatinized Flour by Boiling at 98°C in Aqueous Water System Adding 2 % NaCl with and without 0.05% of SSL

Gelatinized flour in a water system with NaCl (flours CBN1, with SSL, and CBN2, without SSL) were prepared by mixing one part of raw flour and ten part of water (1:10 flour:water, w/w), 2 % of NaCl, and 0.5% of SSL (only for CBN1) in a homogenizer (Cowles Dissolver, Model 1 V-G, Serial LA 266, The Cowles Dissolver Co., Inc., Cayuga, New York, USA), and gelatinized (98°C) for 30 min.

The mixture was dried in a drum drier (Sterling Speedtrol, Model 20, serial 810, Sterling Power System, Inc., Indiana, USA) at 152°C (60psi) during 9sec, and milled in the hammer mill (Fitz Mill Comminuting Machine, Model D, The Fitzpatrick Company; Inc., Chicago, USA) with a 00 sieve (0.5mm).

2.2.7. Gelatinized Flour by Autoclaving with and without SSL

Gelatinized flours (CA1 with SSL, and CA2 without SSL) were prepared by mixing one part of raw flour and three parts of water (1:3 flour:water), and 0.5% of SSL (only for CA1) and gelatinized under pressure in a sterilizer (Type QES-20-36-CSI, Serial283674, American Sterilizer, Co., Erie, PA, USA) at 121°C (15psi) during 30min. The mixture was dried in a drum drier (Sterling Speedtrol, Model 20, serial 810, Sterling Power System, Inc., Indiana, USA) at 152°C (60psi) during 9 sec, and milled in the hammer mill (Fitz Mill Comminuting Machine, Model D, The Fitzpatrick Company; Inc., Chicago, USA) with a 00 sieve (0.5mm).

2.2.8. Gelatinized Flour by Enzymatic Dextrinization with and without SSL

Gelatinized flours (CD1 with SSL, and CD2 without SSL) were prepared by mixing one part of raw flour and three parts of buffer phosphate (1:3 flour: buffer), 0.2% of Termamyl® and 0.5% of SSL (only for CD1) and incubated with agitation at 85°C during 120 min. The mixture was dried in a drum drier (Sterling Speedtrol, Model 20, serial 810, Sterling Power System, Inc., Indiana, USA) at 152°C (60psi) during 9sec, and milled in the above mentioned equipment.

Three replicates (batches) of each type of flour were prepared. All flours were stored at 14 ±2°C in hermetic containers until further analysis.

2.2.9. Analysis of Flours

The following analyses were performed to the raw and processed cassava flours: moisture, crude protein, crude fat, and ash contents (AACC, 2003 methods N°(s): 44-19, 46-13, 30-25, 08-21, respectively). Dietary fiber was determined by the gravimetric enzymatic method (985.29) described in AOAC (1990), using the enzymatic kit from Sigma Chemical No. TDF-100. Total carbohydrate was calculated by differences by using the equation: 100 – (% moisture content + % crude protein content + crude fat content + ash content). Color parameters L, a, b were measured using a Hunter Lab Colorimeter Model D-25; total color change "ΔE", was calculated using the formula: $\Delta E= \sqrt{\Delta L^2+\Delta a^2+\Delta b^2}$ [with the tile standard L: 91.3; a: -0.9 and b: 1.4]. Finally the white index (Hsu et al., 2003) was calculated using the formula: $WI= 100- \sqrt{(100-L^2) + a^2 + b^2}$, and the blue-yellow hue variation with relation to luminosity as L/b (Hunter Lab, 2001).

The particle size distribution was determined in the control and processed cassava flours (Tempranita var.); and it was expressed as the material retained (%) in the following sieves: 20; 35; 42; 60; 80 and 155 mesh (Tyler series), after agitation during 5min in a shaking sieves equipment (Controls, MTR 83111809, Milano, Italy) (AACC, 2003 method N°66-20).

Functional properties of control and processed cassava flours were determined by using the Brabender Farinograph (Haake, Inc, Model 3 300D N° 1832), following the recommended procedures AACC, 2003 method N° 54-21, using 150g (14% moisture basis) for all flours, except for the enzymatic treated flours, were was used 200 g. (14% moisture basis). The parameters evaluated were: absorption (%), arrival time, departure time, time to breakdown, dough development time, mixing tolerance index (MTI), and stability (min). The apparent viscosity of the flour-water system was also evaluated with the amylograph (AACC Method 22-10). Significant point in the paste curve were: initial viscosity, peak viscosity, viscosity at 90 °C, hot-paste viscosity (viscosity after cooking at 90 °C during 30 minutes), and viscosity at 50°C. Three derivatives viscosities amylographic breakdown, setback and consistency were calculated from these basic viscosities.

The ultraestructural features were observed by scanning electron microscopy (SEM). Flour was sprinkled onto pieces of double sided adhesive tape, attached to circular specimen stubs, coated with 50 nm of gold using a Denton vacuum Desk IV Ion Sputter examined at 20.0 kV, and photographed using a JEOL JSM-6490 scanning electron microscope.

2.2.10. Statistical Evaluation of Analytical Data

Data from each flour were analyzed by one way ANOVA followed by the Duncan test, using the Statgraphics® software (Manugistics, 1992). The one way analysis of variance test was utilized to assess significant differences ($P \leq 0.01$ and $P \leq 0.05$) among samples and the Duncan multiple range test was employed to detect which sample(s) was (were) statistically different at the same significance levels.

3. Results and Discussion

3.1. Root Cyanide Concentration

This study is indicating large variations in cyanide content among roots varieties harvested at the 11 months of growing (Table 1).

Table 1. Cyanide concentration (Total, potential and free) of the cassava roots (as dry basis)

Parameters	Tempranita	UCV-2306	UCV-2593	UCV-2192	UCV-2681
Moisture (%)	59.1±0.6	62.3±0.8	58.1±0.3	70.3±0.5	70.3±0.5
HCN Total (mg HCN per 100g dry weight)	<50 *	18.3±0.0	14.2±0.0	123.4±0.0	118.2±0.0
HCN potential (mg HCN per 100g dry weight)	<50 *	14.9±0.0	11.2±0.0	120.1±0.0	115.0±0.0
HCN free (mg HCN per 100g dry weight)	ND	3.4±0.0	3.0±0.0	3.3±0.0	3.2±0.0

* Source: Pérez, et al., (2007).

ND: Non determined

The total cyanide contents in the cassava root varieties were determined and compared. The total cyanide content in root was higher in varieties UCV-2192 (123.4 mg HCN per 100g dry weight) and UCV-2681 (118.2 mg HCN per 100g dry weight) than those shown by the other three that were less than 50 mg HCN per 100g dry weight (Table 1). The varieties tempranita, UCV-2306 and UCV-2593 are classified as sweet and the UCV-2192 and UCV-2681 as bitter.

3.2. Chemical Composition and Physical Properties (Color Profile and Particle Size)

Comparisons of the chemical composition as dry basis, except for moisture of the five raw cassava flours are shown in Table 2. As can seen, the moisture contents have varied slightly in the varieties UCV-2593 (10.6%) and tempranita variety (9.1%). Differences in the moisture content of these flours could be due to its intrinsic structure that could affect the rate of water desorption during the drying process. Moreover it is of consensus that as a sample becomes drier, the remaining water can be tenaciously bound and, consequently, more difficult to remove. However, all of the moisture contents if the flours are among the range generally accepted as safe (Zeleny, 1971). Also, varieties tempranita and UCV-2192 have the higher contents of crude protein (3.5 and 1.9% respectively), and ash (2.0 and 3.4%, respectively), than the other three flours, due to its botanical origin. All of these differences affect the total carbohydrate contents.

Table 2. Chemical composition (as dry basis, except for moisture) of raw cassava flours

Parameters (%)	Tempranita	UCV-2306	UCV-2593	UCV-2192	UCV-2681
Yield	37.2±1.2	38.3±1.2	42.8±1.6	32.9±2.7	33.7±0.6
Moisture	9.1±0.0a	8.5±0.1a	10.6±0.1b	8.1±0.1a	8.9±0.1a
Crude Protein	3.5±0.0a	1.3±0.2b	1.5±0.1b	1.9±0.1b	1.5±0.1b
Crude Fat	0.5±0.1a	0.4±0.1a	0.7±0.1a	1.1±0.1a	0.8±0.1a
Ash	2.0±0.0a	2.5±0.0a	2.7±0.3a	3.4±0.1b	2.6±0.0a
Dietary fiber	2.3±0.0	5.1±0.2	6.1±0.5	5.8±0.2	5.9±0.7
Total Carbohydrates	84.6a	84.6b	81.1a	83.1a	82.9a

Mean values in rows with the same letter are not statistically different using ANOVA and Duncan multiple rank ($\alpha \leq 0.01$). $n \geq 3$.

The crude protein are similar in all of the UCV-varieties, however the tempanito variety has shown the higher content than those with UCV code. Polyorach et al., 2013, have postulated differences in protein content that are ranging from 1-3%. The INN, 1999 reports a mean of 1.1 g/100 g of fresh edible portion of crude protein for Venezuelan cassava roots. Ash content was higher in the UCV-2192 than the other four that have shown similar ash contents. This could be attributed also to its genetic differences.

Table 3 is showing the chemical composition of raw, control and processed cassava (tempanito var.) flours. The flour prepared from the sweet variety tempanito was chosen to be modified, due to its high protein content and its commercial use as compared with the other four. Moisture content was among the range accepted as safe. However, the process conditions affect the moisture content in the flours. It is clear that the moisture content was dramatically affected by the SSL use. Emulsifiers are commonly added to commercial bakery products to improve bread quality and dough handling characteristics (Kohajdová et al., 2009). It can be noted in the Table 3, that the control without SSL (C1) has diminished its moisture content to a half, as compared with the raw flour (C), while it has not happened in the C2. Indeed, all samples processed thermal or enzymatically without SSL, have shown lower moisture content, than those in its counterpart with SSL. It have been postulated that the surfactants interacts with other components of the flour, such as; starch, protein and lipid (Chung and Tsen, 1975; Gómez, et al., 2012; Kurukji et al., 2013) forming a lamellar structure. Additional experiments with waxy maize starch led to the conclusion that SSL impacts swelling power and carbohydrate leaching more by covering the starch granule surface than by forming amylose–lipid complexes (Van Steertegem et al., 2013). Consequently, traditional methods of determination of moisture content, cannot measure the water associated to the lamellar structure or SSL impede the water liberation by covering the starch granules. Emulsifiers can also have an affect on the moisture distribution between the protein and starch fraction of a food system. A decrease in the amount of water imbibed by the starch makes more water available for hydration of the gluten, and this too is thought to contribute to stalling retarding (Kohajdová et al., 2009).

Protein contents were not affected by the treatment used or additive added. While the ash and starch contents were affected by either the treatment and NaCl addition. The ash contents were increased by the use of NaCl and the enzymatic treatment with phosphate buffer, due the uses of salts in both cases, while starch contents were lightly affected when using NaCl, and majorly when using the enzymatic treatment.

Table 3. Chemical composition as dry basis, except for moisture, of control and processed cassava flours

Parameters (%)	Raw : Tempranita var.	Control		Boiling		Boiling with NaCl		Autoclaving		Enzymatic	
	C	C1	C2	CB1	CB2	CBN1	CBN2	CA1	CA2	CD1	CD2
Moisture	9.1±0.0a	4.1±0.2e	9.6±0.2a	3.6±0.2e	8.5±0.2b	3.4±0.1e	6.8±0.0c	5.3±0.1d	7.7±0.2b	3.4±0.1e	4.6±0.1e
Crude protein	3.5±0.0a	3.7±0.1a	3.3±0.2a	3.6±0.6a	3.0±0.4a	3.2±0.1a	3.2±0.2a	3.2±0.3a	3.3±0.1a	3.4±0.1a	3.1±0.0a
Ash	2.0±0.0a	1.9±0.1a	2.0±0.2a	2.3±0.1a	1.9±0.1a	4.04±0.1b	4.0±0.0b	1.8±0.1a	1.8±0.2a	4.7±0.0b	4.4±0.0b
Total Carbohydrates	85,4	90,3	85,1	90,5	86,6	89,36	86	89,7	87,2	88,5	87,9

Mean values in rows with the same letter are not statistically different using ANOVA and Duncan multiple rank ($\alpha \leq 0.01$). $n \geq 3$.

C1 = Control flour; CB1= Gelatinized flour by boiling; CBN1=Gelatinized flour by boiling with NaCl; CA1= Gelatinized flour by autoclaving; CD1= Gelatinized flour by enzymatic dextrinization.

C2 = Control flour with SSL; CB2= Gelatinized flour by boiling with SSL; CBN2=Gelatinized flour by boil with NaCl with SSL; CA2= Gelatinized flour by autoclaving with SSL; CD2= Gelatinized flour by enzymatic dextrinization with SSL.

Table 4. Color profile of the raw cassava flours

Parameters (%)	Tempranita	UCV-2306	UCV-2593	UCV-2192	UCV-2681
Color					
L	89.1±0.0a	98.2±0.0b	97.2±0.0b	95.1±0.0b	97.9±0.1b
a	-1.9±0.0a	0.9±0.0b	0.8±0.0b	0.9±0.0b	2.1±0.0c
b	9.9±0.0a	11.5±0.0b	9.6±0.0a	15.0±0.0c	8.4±0.0a
ΔE	9±0.0a	12±0.0b	10±0.0a	14±0.0c	7±0.0d
WI	85.2±0.2a	88.7±0.0b	89.5±0.3b	89.5±0.0b	90.8±0.0b

Mean values in rows with the same letter are not statistically different using ANOVA and Duncan multiple rank ($\alpha \leq 0.01$). $n \geq 3$. C1 = Control flour with.

Table 4 is showing the color profile of the five flours. The Color Test is a simple test, but it can be very useful to ensure that flour purity remains constant among varieties.

The white index value is demonstrating that the UCV-coded flours are lightly most white the tempanito flour. The ΔE is comparing of the white tile standard with the white color of the flours then it can be see that the whiter as measured comparing a standard white is the UCV-2681 followed by the tempanito flour.

Table 5 resumes the color profile of the control, and modified cassava flours. Since the color axis, b, is positive for yellow colors, the L/b relation in these flours are indicative of the darkness or clarity of the yellow hue of the flour or the presence of pigments. Treatments and additive uses have intensified the yellow hue in all samples. This fact is corroborated by the ΔE values observed by effect of the treatment and use of NaCl and SSL. Indeed, the SSL addition is yielding flour more similar, but usually less white (WI data) than the to the raw flour color.

The particle size of flour is important to producing high-quality food products. For example in elaboration of pasta the particles size of the flour have to be of a size such that no more than 10% will pass through a 180-μm sieve. If the particle size is too fine, the starch leaches into the cooking water and the pasta is mushy. Moreover, Starch-gelatinization temperatures, enzymatic hydrolysis rate, hydration properties and water mobility are strongly correlated with the flour particle size, indicating that those flours with larger particle size have a greater barrier for heat transfer, etc. (Hasjim et al., 2013, De la Hera et al., 2013, Cornejo-Villegas et al., 2013, Blanchard et al., 2012, Iwuoha and Nwakanma, 1998).

Table 5. Color profile of the control, and processed cassava flours

Parameters (%)	Raw : Tempranita var.	Control		Boiling		Boiling with NaCl		Autoclaving		Enzymatic	
	C	C1	C2	CB1	CB2	CBN1	CBN2	CA1	CA2	CD1	CD2
Color											
L	89.1±0.0	87.5±0.2	86.4±0.3	88.4±0.2	86.3±0.6	89.2±0.1	88.5±0.0	86.2±0.2	83.5±0.1	86.7±0.1	86.3±0.1
a	-1.9±0.0a	1.5±0.0	0.8±0.0	1.8±0.2	0.5±0.2	1.4±0.1	0.8±0.0	0.4±0.2	0.4±0.1	0.4±0.1	0.8±0.1
b	9.9±0.0a	12.5±0.2	14.6±0.1	11.2±0.2	11.6±0.3	9.9±0.1	12.9±0.0	14.7±0.2	14.9±0.1	12.9±0.1	13.1±0.1
L/b	9.0±0.0a	7.0±0.1c	5.5±0.2b	7.9±0.4c	7.5±0.6c	9.0±0.1d	6.7±0.2b	5.6±0.1a	5.9±0.3a	6.7±0.0a	6.6±0.1a
ΔE	_	3.1±0.1	5.5±0.5	1.4±0.1	3.5±0.1	0.5±0.1	3.2±0.0	5.8±0.2	7.3±0.1	4.1±0.0	4.4±0.0
WI	85.2±0.2a	82.3±0.1c	80.0±0.2c	83.8±0.1c	82.0±3.4c	85.3±0.3b	82.7±0.1c	79.8±0.2c	77.8±0.1b	81.5±0.0a	81.0±0.1a

Mean values in rows with the same letter are not statistically different using ANOVA and Duncan multiple rank ($\alpha \leq 0.01$). $n \geq 3$. C1 = Control flour with.

C1 = Control flour; CB1= Gelatinized flour by boiling; CBN1=Gelatinized flour by boiling with NaCl; CA1= Gelatinized flour by autoclaving; CD1= Gelatinized flour by enzymatic dextrinization.

C2 = Control flour with SSL; CB2= Gelatinized flour by boiling with SSL; CBN2=Gelatinized flour by boil with NaCl with SSL; CA2= Gelatinized flour by autoclaving with SSL; CD2= Gelatinized flour by enzymatic dextrinization with SSL.

The CODEX STAN 152-1985 for wheat flour required that the 98% or more of flour shall pass through a 212 micron (No. 70 sieve). The granular size distribution of the different flours here studied varied from less than 355 to more than 125 microns, with a particle size population most concentrated between 180 to 125 microns (Figure 1). The thermal treatments reduce the granular sizes in the population and it is with most effectiveness when using SSL. However, none of the flours is according to the CODEX STAN 152-1985.

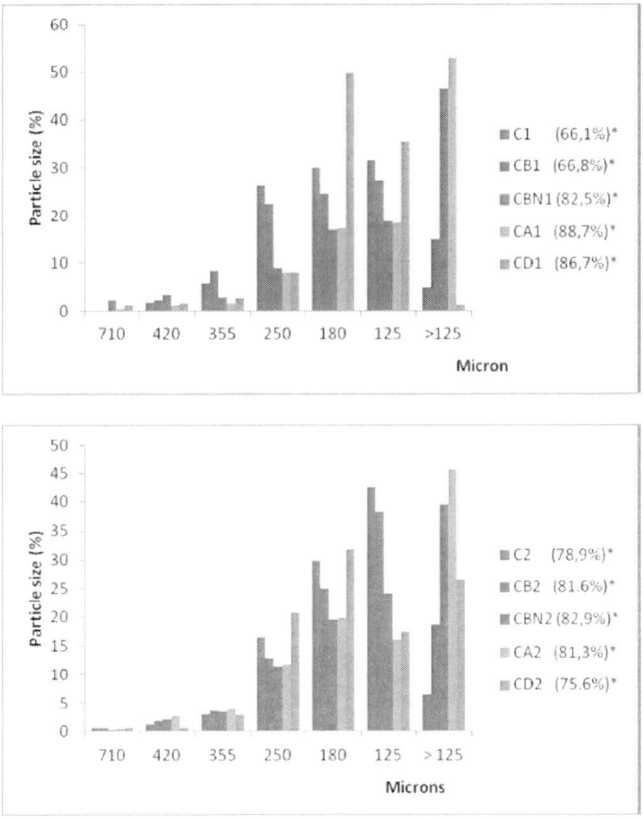

Figure 2. Particle size distribution of the control and modified cassava flours. C1 = Control flour; CB1= Gelatinized flour by boiling; CBN1=Gelatinized flour by boiling with NaCl; CA1= Gelatinized flour by autoclaving; CD1= Gelatinized flour by enzymatic dextrinization. C2 = Control flour with SSL; CB2= Gelatinized flour by boiling with SSL; CBN2=Gelatinized flour by boil with NaCl with SSL; CA2= Gelatinized flour by autoclaving with SSL; CD2= Gelatinized flour by enzymatic dextrinization with SSL. * % of flour that is passing through a 212 microns.

3.3. Rheological and Ultraestructural Properties

As can see in Figure 3, thermal treatments have changed drastically the viscosity behavior of the flour as compared with those shown for the raw flour. The flours just dry (C1), boiled, without (CB1) and with NaCl (CBN1), and autoclaved (CA1) have shown a high initial viscosity. This vicosity diminished in the holding and cooling period at the raw, C1, and CB1 flours. While CBN1 and CA1 remaining relatively high. Except for the flours dextrinized, the SSL addition largely modified the amylograph parameters of all flours, making them similar during the holding and cooling time. SSL also have lightly diminished the retrogradation tendency in the flours. Dextrinized flour with and without SSL did not develop viscosity at 7%, showing a line close to the cero en the amylographic plot (Figure 3). Figure 4, shown the response of each flour when were analyzed by using the Farinograph Brabender. The first phase of mixing is introduction of water to perform the dough. Water is absorbed to reach the top (500 BU) defining the arrival time. As can be noted the water the farinograph mixing curve showed a rise in water absorption and a decrease in dough development time with greater weakening, when SSL is added. On the other hand, when comparing the result of the water absorption of the cassava flour (Figure 3), with those of the wheat flours; which is varying from 63-75% (Shuey 1972) as a function of the protein content, it is a big contrast. These results are in agreement with the high viscosity that is observed in the flour when it is thermally modified. Except in the enzymatic treatment where the effect is inverse when using SSL, the arrival time has delayed in a fifty percent.

The dough stability – is the difference in time, to the nearest half-minute, between the time when the curve first intercepts the 500 BU line (arrival time) and the time when the curve leaves the 500 BU line (departure time). It is generally accepted that longer stability of flour, the more tolerance is to mixing (D'Appolonia and Kunerth, 1998). This parameter is quite affected by the use of SSL in all of the treatments. The Mixing Tolerance Index (MTI) is also augmented by the use of SSL indicative of diminution of the tolerance to the mix.

Flours with low MTI have a good tolerance to mixing, while flours with high MTI's are critical to mixing and especially to overmixing (Shuey 1972). The Figure 3 is showing, SSL on the cassava flour is working in an inverse mode, as compared to those that have been reported in literature for wheat flours (Orthoefer, 2010, Azizi and Rao, 2004, Indrani and Rao, 2003; Taha, 1990).

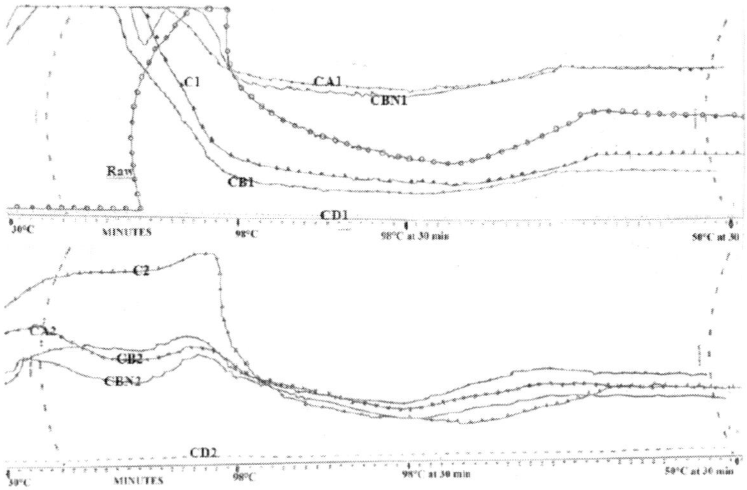

Figure 3. Amylographic profile of the flours elaborated by thermal and enzymatic treatments. C1 = Control flour; CB1= Gelatinized flour by boiling; CBN1=Gelatinized flour by boiling with NaCl; CA1= Gelatinized flour by autoclaving; CD1= Gelatinized flour by enzymatic dextrinization. C2 = Control flour with SSL; CB2= Gelatinized flour by boiling with SSL; CBN2=Gelatinized flour by boil with NaCl with SSL; CA2= Gelatinized flour by autoclaving with SSL; CD2= Gelatinized flour by enzymatic dextrinization with SSL.

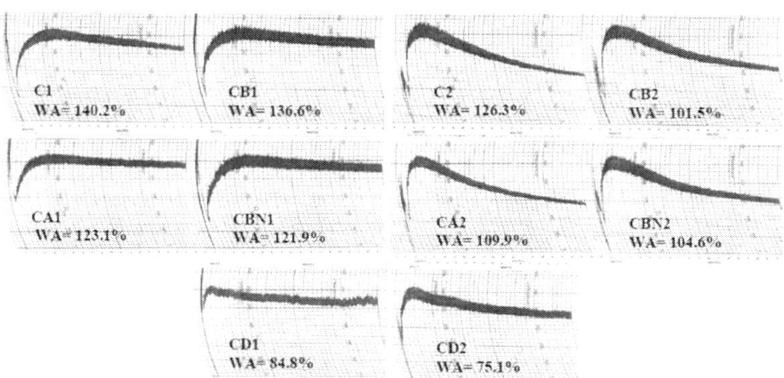

Figure 4. Farinograph profile of the flours elaborated by thermal and enzymatic treatments. WA: Water Absorption. C1 = Control flour; CB1= Gelatinized flour by boiling; CBN1=Gelatinized flour by boiling with NaCl; CA1= Gelatinized flour by autoclaving; CD1= Gelatinized flour by enzymatic dextrinization. C2 = Control flour with SSL; CB2= Gelatinized flour by boiling with SSL; CBN2=Gelatinized flour by boil with NaCl with SSL; CA2= Gelatinized flour by autoclaving with SSL; CD2= Gelatinized flour by enzymatic dextrinization with SSL.

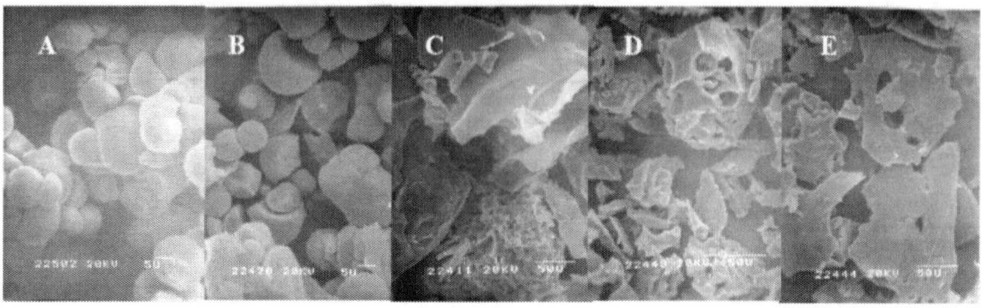

Figure 5. Gelatinization of the granular starch population, as compared to those presented of cassava starch and raw cassava flour. A: Cassava starch. B: Cassava raw flour. C: Gelatinized cassava flour by boiling. D: Gelatinized cassava flour by autoclaving. E: Gelatinized cassava flour by enzymatic dextrinization. The counterparts of each treatment with and without NaCl, and SSL have shown are same gelatinization behaviors.

The different cassava flours without SSL are showing curve type III and IV, with medium peak time and long stability good flour for bakery due they have good elastic properties; while those with SSL shown curves type I with short peak time and short stability. These flours are adequate for cake, cookies, and cracker manufacture (Shuey 1972).

Figure 5 reveal the total gelatinization of the flours by effect of the treatments as compared with the granular of the cassava pure starch, and raw flour.

The counterparts of each treatment with and without NaCl, and SSL have shown are same gelatinization behaviors. It can be noted that the cassava raw flour was dried at 45 °C (Pérez et al., 2007) to avoid the starch gelatinization, and in the figure 5B is corroborating that the starch granular structure was not affected for the drying temperature.

Changes in starch are function of both temperature, and amount of available water. If temperature is maintained at 100 ªC, the changes in starch depend principally on the amount of available water (Derby et al., 1975). Derby corroborated a fully gelatinization (as in gravy) with 60% of moisture or with 17:25 starch (db)/ water ratio.

Indeed, Derby et al., 1975 have established that the first observed change in structure where a 50 % of loss of birefringence occurred, considered as the first stage of gelatinization or point of gelatinization was with 33 % of water or with 52:25 starch (db)/ water ratio. In the study the amounts of water added at the different treatments were superior to those pointed out for Derby et al., 1975, and the effect was corroborated by the total gelatinization observed by SEM on the flours.

Conclusion

Thermal and enzymatic methods with and without SSL, and NaCl of modification, applied in this study to cassava flour are causing modifications in their physical, chemical and rheological properties. Major changes have occurred in its granular sizes, color and, especially, in the rheological (farinographic profile) properties of them. One of the most significant was the SSL addition; which produce changes on the farinographic profile of the cassava flours in an inverse way to those reported in literature for wheat flour. The specific functional properties of thermal and enzymatic processed cassava flours could be functionally adequate for their incorporation into products formulation such as, for instance, baked goods and breads.

References

AACC American Association of Cereal Chemist. (2003). *Laboratory Method.* Ed. St. Paul. MN, USA.

AOAC (1990). *Official methods of Analysis of the Association of the Official Analytical Chemists.* 15 ed., Washington. U.S.A.

Aryee, F.N.A.; Oduro, I., Ellis, W.O., & Afuakwa, J.J. (2006). The physicochemical properties of flour samples from the roots of 31 varieties of cassava. *Food Control*, 17(11): 916-922.

Azizi, M.H., & Rao, G.V. (2004). Effect of surfactant gels on dough rheological characteristics and quality of bread. *Critical Reviews in Food Science and Nutrition*, 44(7-8), 545-52.

Banea, J.P., Bradbury, J.H., Mandombi, C., Nahimana, D., Denton, I.C., Kuwa, N., & Katumbay D.T. (2013). Control of konzo by detoxification of cassava flour in three villages in the Democratic Republic of Congo. *Food and Chemical Toxicology*, 60:506-513.

Blanchard, C., Labouré, H., Verel, A., & Champion, D. (2012). Study of the impact of wheat flour type, flour particle size and protein content in a cake-like dough: Proton mobility and rheological properties assessment. *Journal of Cereal Science*, 56 (3):691-698.

Bradbury, J.H., & Denton, I.C. (2010). Rapid wetting method to reduce cyanogen content of cassava flour. *Food Chemistry*, 121(2):591-594.

Cardoso, A.P., Mirione, E., Ernesto, M., Massaza, F., Cliff, J., Haque, M.R., & Bradbury, J.H. (2005). Processing of cassava roots to remove cyanogens. *Journal of Food Composition and Analysis*, 18(5): 451-460.

Casier, J.P.J., De Paepe, G.M.J., Willems, H.E.J., Goffings, G.J.G., Hermans, J.L, & Noppen, H.E. (1979). Bread production from pure flours of tropical starchy crops: iii from pure and mixed flours of cassava, millet, sorghum, corn, rice, and the starches *Tropical Food: Chemistry and Nutrition*, 279-340.

Charles, A.L., Huang, T.C., Lai, P.Y., Chen, C.C.., Lee, P.P., & Chang Y.H. (2007). Study of wheat flour–cassava starch composite mix and the function of cassava mucilage in Chinese noodles. *Food Hydrocolloids*, 21(3): 368-378.

Charoenkul, N., Uttapap, D., Pathipanawat, W., & Takeda Y. (2011). Physicochemical characteristics of starches and flours from cassava varieties having different cooked root textures. *LWT - Food Science and Technology*, 44 (8):1774-178.

Chung, O.K., & Tsen, C.C. (1975). Functional properties of surfactants in bread making. I. Role of surfactants in relation to flours constituents in a dough system. *Cereal Chemistry* 52:832-843.

Chung, O.K., & Tsen, C.C. (1977). Functional properties of surfactants in bread making .II. Compsoition of lipids associated with dough containing various levels of surfactants. *Cereal Chemistry* 54:857-864.

Ciarfella, A.T., Pérez, E., Tovar, J., Sánchez, T., & Dufour, D. (2013). Efecto de la adición de harina de quinchoncho (*Cajanus cajan*) sobre la calidad nutricional del casabe. *Revista de la Facultad de Agronomía de la Universidad del Zulia*. 30(1):131-148.

Codex Standard for Wheat Flour. (1985). *Codex Stan*-152-1985. In: www.codexalimentarius.org/input/download/standards/.../CXS_152e.pdf.

Cornejo-Villegas, M.A., Gutiérrez-Cortez, E., Rojas-Molina, I., Del Real-López, A., Zambrano-Zaragoza, M.L., Martínez-Vega, V., & Rodríguez-García M.E. (2013). Physicochemical, morphological, and pasting properties of nixtamalized flours from quality protein maize and its particle distribution. *LWT - Food Science and Technology*, 53 (1): 81-87.

Cooke, R. (1978). An enzymatic assay for the total cyanide content of cassava (*Manihot esculenta Crantz*). *Journal of Science and Food Agriculture*, 29: 345- 352.

Cueto Bautista, D., Pérez Sira, E., & Dufour, D. (2012). Formulación de mezcla para elaborar torta para regímenes especiales. [Formulation of mixture for making cake for special regimes]. *Memorias del VIII*

Congreso Iberoamericano de Ingeniería de Alimentos CIBIA 8, Lima Perú.

Cumbana, A., Mirione, E., Cliff, E.J., & Bradbury J.H. (2007). Reduction of cyanide content of cassava flour in Mozambique by the wetting method. *Food Chemistry*, 101 (3): 894-897.

Essers S., Bosveld M., van der Grift R., & Voragen A. (1993). Studies on the quantification of specific cyanogens in cassava products and introduction of a new chromogen. *Journal of Science and Food Agriculture*. 63: 287–296.

D'Appolonia B.L. & Kunerth WH. (1998). *Farinograph Handbook. A practical guide to using the Farinograph* (3er ed.) Edited by Bert L. D'Appolonia and Wallace H. Kunerth American Association of Cereal Chemistry (AACC).

De la Hera, E., Gómez, M., & Rosell, C.M. (2013). Particle size_distribution of rice flour affecting the starch enzymatic hydrolysis and hydration properties. *Carbohydrate Polymers*, 98 (1):421-427.

Derby, R.I., Miller, B.S., Miller, B.F., & Trimbo H.B. (1975). Visual Observation of Wheat-Starch Gelatinization in Limited Water Systems. *Cereal Chemistry*, 52 (5):702 - 713.

FAO. 1999. Raíces tubérculos plátanos y bananas en la nutrición humana. Organización de las Naciones Unidas para la Agricultura y la Alimentación. *Colección FAO: alimentación y nutrición*. Roma.

Fiorda, F.A., Soares, Jr. M.S., Da Silva, F.A., Grossman, M.V.E., & Souto, L.R.F. (2013). Microestructure, texture and colour of gluten-free pasta made with amaranth flour, cassava starch and cassava bagasse. *LWT - Food Science and Technology*, 54 (1):132-138.

FFC Food Chemical Codex. (1981). *Sodium Stearoyl Lactylate* (7 ed.). National Academic Press Washington, DC. pp. 964–965.

Gómez, A.V., Ferrer, E., Añón, M.C., & Puppo M.C. (2012). Analysis of soluble proteins/aggregates derived from gluten-emulsifiers systems. *Food Research International*, 46 (1):62-68.

Hasjim, J., Li, E., & Dhital S. (2013). Milling of rice grains: Effects of starch/flour_structures on gelatinization and pasting properties. *Carbohydrate Polymers*, 92(1): 682-690.

Hsu C-L., Chen, W., Wenga, Y-M., & Tsenga C-Y. (2003). Chemical composition, physical properties, and antioxidant activities of yam flours as affected by different drying methods. *Food Chemistry*, 83: 85-92.

Hunter Lab Manual. 2001. *Hunter associates laboratory universal software version* 3.8. ISO 9001 Certified. 55p.

Indrani, D, & Rao, G.V. (2003). Influence of surfactants on rheological characteristics of dough and quality of parotta. *International Journal of Food Science & Technology*, 38 (1):47-54.

INN Instituto Nacional de Nutrición (1999). *Tabla de Composición de Alimentos* (TCA) Pub 52 Serie de Cuadernos Azules. Caracas- Venezuela.

Iwuoha, C.I., & Nwakanma, M.I. (1998). Density and viscosity of cold flour pastes of cassava (*Manihot esculenta Crantz*), sweet potato (*Ipomoea batatas* L. Lam) and white yam (*Dioscorea rotundata* Poir) tubers as affected by concentration and particle size. *Carbohydrate Polymers*, 37 (1): 97-101.

Jisha, S., Padmaja, G., Moorthy, S.N., & Rajeshkumar K. (2008). Pre-treatment effect on the nutritional and functional properties of selected cassava-based composite flours. *Innovative Food Science & Emerging Technologies*, 9 (4): 587-592.

Khalil, A.H., Mansour, E.H., & Dawoud F.M. (2000). Influence of Malt on Rheological and Baking Properties of Wheat–cassava Composite flours *LWT - Food Science and Technology*, 33(3): 159-164.

Kohajdová, Z., Karovičová, J., & Schmidt, Š. (2009). Significance of emulsifiers and hydrocolloids in bakery industry. *Acta Chimica Slovaca*, 2 (1):46 – 61.

Kurukji, D., Pichot, R., Spyropoulos, F., & Norton I.T. (2013). Interfacial behaviour of sodium stearoyllactylate (SSL) as an oil-in-water pickering emulsion stabilizer. *Journal of Colloid and Interface Science*, 409 (1): 88-97.

Kwee, H.W, Sidwell, V., Wiley, R., & Hammerle, O.A. (1969). Quality and nutritive value of paste made from rice, corn, soya and tapioca enriched with fish protein concentrate. *Cereal Chemistry*, 47, (1):78-75.

Manugistics. (1992). Statgraphics reference manual version 6. *Manugistics.* Inc. Maryland.

Mark, E., Ukhun, M.E., & Nkwocha, F.O. (1989). The hydrocyanic acid (HCN) content of *garri flour* made from cassava (*Manihot* spp.) and the influence of length of fermentation and location of source. *Food Chemistry*, 33 (2): 107-113.

Muzanila, Y.C., Brennan, J.G., & King R.D. (2000). Residual cyanogens, chemical composition and aflatoxins in in cassava flour from Tanzanian villages. *Food Chemistry*, 70(1): 45-49.

Nambisan, B. (2011). Strategies for elimination of cyanogens from cassava for reducing toxicity and improving food safety. *Food and Chemical Toxicology* 49(3): 690-693.

Nindjin, C., Amani, G.N., & Sindic M. (2011). Effect of blend levels on composite wheat doughs performance made from yam and cassava native starches and bread quality. *Carbohydrate Polymers*, 86(4): 1637-1645.

Numfor, F.A., Walter Jr., W.M., & Schwartz, S.J. (1998). Emulsifiers affect the texture of pastes made from fermented and non-fermented cassava flours. *International Journal of Food Science and Technology*, 33 (5): 455–460.

Oboh, G., & Akindahunsi, A.A. (2005). Nutritional and toxicological evaluation of *Saccharomyces cerevisae* fermented cassava flour. *Journal of Food Composition and Analysis*, 18(7):731-738.

Orthoefer, F. (2010). Applications of Emulsifiers in Baked Foods. In Hasenhettl, G.L.; Hartel, R.W. *Food Emulsifiers and Their Applications* (2 ed.). New York: Springer. pp. 263–284.

Osman, E.M. (1975). Interaction of starch with other components of food systems. *Food Technology*, 29(4):30-44.

Osman, E.M. Leith SJ, & Flesh M. (1961). Complexes of amylose with surfactants. *Cereal Chemistry* 38:449-462.

Ouraga-Djoussou L.H. & Bukanga M. (1998) Cassava and wheat consumption in Africa: New aooprtunieties for cassava in the 21-st century. In Akorada, A.O. and Ekanayake, I.J. (eds) *Proceeding of the 6th Symposium if the International Society for Tropical Root Crops- Africa Branch.* International Society of Tropical Root and Tuber Crops- Africa Bramnch-Idaban Nigeria, pp 328-332.

Pasqualone, A., Caponio, F., Summo, C., Paradiso, V.M., Bottega, G., & Pagan, M.A. (2010) Gluten-free bread making trials from cassava (*Manihot esculenta Crantz*) flour and sensory evaluation of the final product. *International Journal of Food Properties* 13: 3, 562 — 573.

Pérez, E., Lares, M., González, Z., & Tovar J. (2007). Production and characterization of cassava (*Manihot esculenta* Crantz) flours using different thermal treatments. *INTERCIENCIA*, 32 (9):615-619.

Polyorach S., Wanapat, M., & Wanapat, S. (2013). Enrichment of protein content in cassava (*Manihot esculenta* Crantz) by supplementing with yeast for use as animal feed. *Emirates Journal of Food and Agriculture*. 25 (2): 142-149.

Santana MA, Vásquez V, Matehus J, & Aldao RR. (2002). Linamarase expression in cassava cultivars with roots of low- and high-cyanide content. *Plant Physiology*, 129(4):1686-1694.

Shittu, T.A., Sanni, L.O., Awonorin, S.O., Maziya-Dixon, B., & Dixon A. (2007a). Use of multivariate techniques in studying the flour making properties of some CMD resistant cassava clones. *Food Chemistry*, 101(4):1606-1610.

Shittu, T.A., Raji, A.O., & Sanni L.O. (2007b). Bread from composite cassava-wheat flour: I. Effect of baking time and temperature on some physical properties of bread loaf. *Food Research International*, 40(2):280-290.

Shittu, T.A., Dixon, A., Awonorin, S.O., Sanni, L.O., & Maziya-Dixon B. (2008). Bread from composite cassava-wheat flour—II. Effect of cassava genotype and nitrogen fertilizer on bread quality. *Food Research International*, 41(6):569–578.

Shittu, T.A., Aminu, R.A., & Abulude, E.O. (2009). Functional effects of xanthan gum on composite cassava-wheat dough and bread. *Food Hydrocolloids*, 23 (8): 2254–2260.

Shuey W.C. (1972). The Farinograph Handbook, ed. W.C Shuey, *American Association of Cereal Chemists* (AACC) pp 20-32.

Stampfli, L., & Nersten, B. (1995). Emulsifiers in bread making. *Food Chemistry* 52 (4): 353-360.

Taha, S. A. (1990). Effect of sodium stearoyl-2-lactylate on some quality parameters of pasta made from *durum* wheat semolina. *Cereal Research Communications*, 18 (4): 321-327.

Tivana, L.D., Dejmek, P., & Bergenståhl, B. (2010). Characterization of the agglomeration of roasted shredded cassava (Manihot esculenta crantz) roots," *Starch/Staerke*, 62 (12): 637– 646.

Van Steertegem, B., Pareyt, B., Brijs, K., & Delcour, J.A. (2013). Combined impact of *Bacillus stearothermophilus* maltogenic alpha-amylase and surfactants on starch pasting and gelation properties. *Food Chemistry*, 139(1–4):1113-1120.

Westby, A. (2002). Cassava utilization, storage and small-scale processing. In *Cassava: Biology, production and Utilization*, R. J. Hillocks, J. M. Thresh, and A. C. Belloti, Eds., pp. 281–300, CAB International, 2002.

Wang, Y., Zhang, M., & Mujumdar, A.S. (2012). Influence of green banana flour substitution for cassava starch on the nutrition, color, texture and sensory quality in two types of snacks *LWT - Food Science and Technology*, 47(1):175-182.

Wootton, M., & Bamunuarachchi, A. (1980). Application of differential scanning calorimetry to starch gelatinization. III. Effect of sucrose and sodium chloride. *Starch–Stärke*, 32(4):126-129.

Zeleny, L 1971. Criteria of wheat quality: In: Pomeranz, J. (Ed.). Wheat Chemistry and Tecnology. Pub. *American Association of Cereal Chemists AACC*, St Paul MN.

In: Cassava
Editor: Francis P. Molinari
ISBN: 978-1-63321-031-8

Chapter 3

PROXIMATE COMPOSITION, MORFOMETRIC AND FUNCTIONAL PROPERTIES OF NATIVE AND MODIFIED CASSAVA STARCHES

L. J. Sívoli[1*], A. T. Ciarfella[2] and E. E. Pérez[3]
[1]Departamento de Ciencias Biomédicas, Facultad de Ciencias Veterinarias, Universidad Central de Venezuela
[2]Unidad de Cursos Básicos, Núcleo de Anzoátegui, Universidad de Oriente
[3]Instituto de Ciencia y Tecnología de Alimentos, Facultad de Ciencias Universidad Central de Venezuela

ABSTRACT

Native cassava starch have adequate functional properties; for example, their swollen quickly to low temperatures producing clear, sticky and fibrous pastes, that make them useful for their use in different industrial applications. As a contrast, they also have some limitations in their properties; such as, low stability to high temperatures (90°C), to pH changes, and to mechanical process, that often are restrictive their industrial use. These limitations in its functional properties are improved and optimized by the modification chemical and/or physical of this starch. The goals of this study were to evaluate the proximate composition, physical characteristics, morphometric parameters and functional properties of starches isolated and purified at the laboratory from cassava clones of the Venezuelan Germplasm Bank, and those from

[*] Corresponding author: Email: sivoli.lilliam@gmail.com.

commercial cassava starches (native and crosslinked) acquired in the local market, and to compare all of these properties with those of its modified counterpart by using the following methodologies: physical (extrusion, dehydration in double drum, gamma irradiation and microwaves) and chemical (crosslinked, crosslinked-acetylated). It was also assayed, at a pilot level, critical parameters of the industrial process of crosslinked-acetylated modification in order to optimize it. Results have shown starches with high purity, and with a composition lightly affected by the modification treatments used, with the exception by those irradiated, where an increment of the contents of total and reducing sugar, and a diminution of its pHs were observed. Color of the starches was affected by the physical methods, making their less white. The contents of amylose apparent of the chemical modified starches have slightly augmented to 35.1-36.6%; as compared with its counterpart commercial native (34.13%), with a degrees of modification of: 0.013 (cross-linked), 0.035 (acetylated alone) and 0.064 (cross-linked-acetylated). The functional characteristics of starches were affected drastically, as function of the kind and severity of the treatments. The rheological behavior of native cassava starches is of a non-Newtonian pseudo plastic fluid; increasing its viscosity, as a function of the diminution of the temperature and with a pronounced viscosity peak, high granular breakdown, and retrogradation tendency. In the case of modified starches, its overall viscosities, granular breakdown, and retrogradation tendency are diminished with the chemical modification treatments used. Conversely, the extrusion treatments have shown the opposing effects. The drum-dried starch had developed an initial viscosity, without maximum peak and was very stable during the cooking. With the exception of the starch modified by dehydration (drum dried and extrusion); where the grain have shown a partial or total destruction of its granular structure, and loss of its Maltese cross, the other modification treatments are only affecting with a lightly augmentation of its sizes, keeping intact their structure, with its edges affluent defined, and forms three-dimensional, depressions corresponding to the *hilum*, smooth surface, and without the pore presence. The patterns of X-rays diffraction of the starches chemically modified do not vary, when comparing to its native counterpart. As conclusion it was demonstrated the feasibility to improve the cassava starch properties by modification for to extend its use in the different industries.

Keywords: Cassava starches, physical modification, chemical modification, functional properties

INTRODUCTION

Starch, the energetic reserves of the plant cells, is a significant source of carbohydrates in the human diet. Most of the nutritional values of the main human food (potato, rice, maize, wheat, among other) are provide for starch. Although starches from different sources exhibit a similar basic structure, they differ in specific details of your micro and ultra structure. These differences in regards to the granular morphology; such as size, shape, presence of pores, channels and cavities, molecular structure (conformational structure of amylose and amylopectin) (McKee and McKee, 2003; Thomas and Atwell, 1999), crystalline type and degree (Atichokudomchai et al., 2001) and interrelation of amylose/amylopectin (Thomas and Atwell, 1999), with non-starch components (protein, fatty material, mineral) are responsible for the variations in the properties of starch. All of these differences are reflected in the patterns of X-ray diffraction, temperature range of gelatinization, gel properties, tendency to retrogradation, pattern of swelling power, water absorption, solubility of its granules, among the most important; and which are indicative of divergence in the functional properties among starches from various botanical origins (Huber y BeMiller, 2000).

Starch granule is composed mainly of two polysaccharides: amylose and amylopectin (McKee and McKee, 2003). These two components, tied together which and oriented radially, are responsible for forming the semi-crystalline structure of the starch granules, which consists of a crystalline lamellae (ordered, closely packed string of parallel glucan, with amylopectin predominance) structure, and amorphous regions (less ordered regions where predominate branching points, with amylose predominance). Starches from different sources have different patterns and degrees of crystallinity (Atichokudomchai et al., 2001). Starch is the main component of the cassava roots (Beleia et al., 2006; INN, 1999). Cassava starch is commercially one of the important starches for food ingredient, and it is also used paper industry, textile industries in tropical countries (Radley, 1976; Whistler et al., 1984; Nabeshima and Grossmamm, 2001). The cassava starch granules in its native state possess a several important properties; for example, fast swelling a low temperature, providing clear, sticky, stringy pastas (Whistler et al., 1984). These properties make it useful in limited industrial applications; all the more, it has negative characteristics; for example its low stability to: high temperatures (above 90 °C), low pH, and action of mechanical forces; which make it instable and non-useful for several other processes. To improve these dysfunctional properties of the cassava starch, there are industrial procedures;

such as, the chemical modifications. Among these modifications are usually applied the cross-linking, esterifications, or combination of this two; also called derivation (Wurzburg, 1986). Another type of modification that also used commonly by the industry is the physical; mainly represented by application of heat (in limited or excessive aqueous system) or irradiation. Each one of them will be used depending on the desired application by food processors to meet specific needs. Even more, the diversity of the modern food industry with its huge variety of food products is requiring that the starch should be able to tolerate a wide range of processing techniques as well as different conditions of distribution, storage and preparation. These demands are met by the modified starches by physical and chemical methods. Indeed, by using these procedures can be optimized several of its properties, and adding another one, for example for to develop low-calorie products, with optimal consistency and palatability, or a myriad of products in different industries (Slaterry et al., 2000; Liu et al., 1999; Nabeshima and Grossmann, 2001; Jyothi et al., 2006). Several techniques have been implemented in order to understand extensively the starch micro structural behavior, and for hence apply most appropriately the starch processing in the production or develop different products including those for food product. Among the employed, it can be mentioned: Differential scanning calorimetry, light and electron microscopy, X-ray scattering and biochemical techniques. These techniques have increased their accuracy and precision each year to provide valuable information for the characterization of the two major polysaccharides the amylose and amylopectin (fine structure), relating them with the functional properties of the modified starches

Chemical modification of starch changes its functionality. The alteration in these properties due to the chemical modification is dependent on the type of modification and the degree of substitution (Engelmann et al., 2001; Choi and Kerr, 2002). Examples; the crosslinked starches have increase their stability during cooking, especially under shear force and low acidity. However, the crosslinking reduces the pasta clarity and stability under cold storage conditions. Moreover, modifications such as acetylating and hydroxypropylation are used to enhance those starch characteristics, that crosslinking could not reach. The used chemicals have been approved by regulatory agencies in the United States to produce modified starches for food uses (Thomas and Atwell, 1999). The chemistry involved in the modification is quite simple in alkaline condition, and mainly are reactions associated with the hydroxyl groups of polymer. Moreover, Fortuna et al., (2000), cites that the susceptibility of starch to chemical modification depends on the dimensions of the granule structure and botanical origin. The authors have reported that small granules being more resistant to external influences and less susceptible to the transformation than the large ones.

The goals of this study were to compile the data of research performed on cassava starch in Venezuela, showing the evaluation of the proximate composition, physical characteristics, morphometric parameters and functional properties of starches isolated and purified at the laboratory from cassava clones of the Venezuelan Germplasm Bank, and those from commercial (native and crosslinked) cassava starches acquired in the local market, and to compare all of these properties with those of the starches modified in the laboratory by using the following methodologies: physical (extrusion, dehydration in double drum, gamma irradiation and microwaves), and chemical (crosslinked, crosslinked-acetylated). It was also assayed, at a pilot level, critical parameters of the industrial process of crosslinked-acetylated modification in order to optimize it.

Material and Methods

Materials

Starches Isolated and Purified from Cassava Clones of the Venezuelan Germplasm Bank

Native starch of four clones (two sweet, and two bitter) harvested from the national collection of the Facultad de Agronomía de la Universidad Central de Venezuela, located in Maracay-Aragua, 10 ° 17 'N and 67 3 `O, height of 445 m and average annual rainfall of 982 mm were isolated and purified by Ciarfella, in 2009 using the procedure described by Pérez et al., 2005. The clones studied were coded as: 2192, 2681 (bitter) and 2306, 2593 (sweet). The origin of these clones of cassava is as follows: CLONE 2192. San José de Costa Rica; CLONE 2681: Maturín, Monagas; CLONE 2306: Santa Bárbara de Barinas and CLONE 2593: River Ajuaro, Guárico.

Commercial Native, Physical and Chemical Modified Starches

Commercial cassava starch, purchased at a local market, another sample supplied by the Company Alfonzo Rivas & Co., La Encrucijada, Turmero, edo. Aragua, Venezuela.

Methods

Modification Procedures of the Commercial Starch

Physical Modification

Extrusion. Three batches of commercial cassava starch were separated. Two of them were conditioned to 25% moisture, leaving the third with its initial moisture content (10.21%). The two batches of samples with 25% moisture were extruded at 150 °C, one in a single screw extruder and the second one in a twin screw extruder (Rheocord torque Rheometer, model 104, and model 300). The third batch was extruded at 100 °C in a single twin screw extruder. The screw speed in all cases was 90 rpm, once modified starches were ground and sieved (60 mesh) and stored in an airtight container for further analysis (Pérez, 1996).

Gelatinization in a drum dryer. A starch-water suspension 1:2.5 (w/w), as recommended by Colonna et al., (1984) and Doublier et al., (1986) of the commercial cassava starch was prepared and dehydrated by using a double drum dryer (Sterling Speedtrol, model 20, serial 810, Sterling Power System, Inc., Indiana, USA) for 9 seconds at 152 ° C. The gelatinized starch was obtained as flakes or thin flakes and ground to achieve a uniform, and fine particle size (60 mesh) and stored in an airtight container for further analysis.

Gamma Irradiation. Commercial cassava starch was irradiated by using a cobalt-60 source, located at the Venezuelan Institute for Scientific Research (IVIC). Processes of dosimetry were conducted in order to quantify the dose rate and the distribution of the incident rays. Samples were packed in 300cm3 flasks and doses of 0.5 and 2.5 Mrad was applied on the sample with initial moisture content (11.60%) at room temperature (Pérez, 1996).

Microwave Irradiation. Commercial cassava starch was conditioned to 25% moisture, as reported by Lewandowicz et al., (1997), and transferred to a polyethylene bag. The polyethylene bag with sample was microwaved in a microwave oven Sanyo brand (20 V, 60 Hz / AC, model EM-370T, chamber measured of: 25.4 cm high, 50.8 cm wide and 40.6 cm deep). The process conditions were: 85 °C, 6 minutes, 50% of power of 650 W and 2450 MHz. The control of temperature was performed by using a thermocouple which was introduced into the polyethylene bag. The microwaved starch was dehydrated at 45 °C for 24 hours in a conventional tray dryer with forced air (National Drying Machinery, TY244864, Philadelphia, Pennsylvania, USA), grounded to a particle size of 0.250 mm, and store in a sealed plastic containers until further analysis.

Chemical Modification

Crosslinked. The native starch was phosphated with $Na_3(PO_3)_3$ at alkaline pH, using the methodology described by Paschall (1964). The DS of the chemically modified starch was determined following methodologies described by Paschall (1964) and was calculated using the equation DS of monosodium esters= $162P/(3100\text{-}102P)$.

Where: P is the difference in percent phosphorus (db) between chemically modified and native starches expressed as percent phosphorus (db).

Crosslinked-Acetylated (double derivated-PAS). Native cassava starch was double derived, in alkaline condition, as follow: first by phosphatation with sodium tripolyphosphate (TPS) and then by acetylation with acetic anhydride.

Native cassava starch was also double derived, in alkaline condition, as follow: first by phosphatation with trimetaphosphate (TMP) (TPS) and then by acetylation with vinyl acetate monomer (VAM).

The substitution level of the modified starch was determined following the methodologies described below.

% Acetyl= [mL (blank) – mL (sample) x N x 0,043 x 100]/
sample weight (db)

DS = [162 x %Acetyl] / [4,300- (42 x % Acetyl)] (Lawal O., 2004).

Test at pilot scale (To optimize the industrial process of double derivation of the Cross linking/Acetylation)

In order to optimize the double derivation of the Cross linking/Acetylation for food industry, a pilot tank of capacity of 500 liters, with the same characteristics of operational tanks or reactors where takes place the process was designed (Sívoli, L., 2009).

Several runs were assayed at the pilot reactor previously designed, evaluating parameters such as: pH (as modified by the amount of sodium hydroxide added to the process), the amount of sodium trimetaphosphate and temperature. The comparison with a control has allowed demonstrating, which kind of changes take place in the gelatinization profile during the modification process, and they are mainly measured with the Brabender viscosity. These parameters are cited in the literature as major modifiers of the process (Sing et al., 2004; Liu et al., 1999, Lim and Sieb, 1993). This predicts or defines the time for cutting out the reaction of modification. Independent variables: concentration of sodium trimetaphosphate, sodium hydroxide concentration and temperature of the modification process. Brabender viscosity in the cutoff modification inbreeding: Dependent Variables.

STARCH CHARACTERIZATION

The native starches experimental clones, native commercial, and physical and chemical modified starches were analyzed as follow:

Yield of the starch obtained from each batch was calculated using the equation:

% yield = (wt of starch isolated/wt of edible portion of the tuber) x 100.

The contents of moisture, ash, crude protein (% N x 6.25), crude fiber, reducing sugars and non-reducing sugars were determined according to official methods 44-19, 08-12, 46-13, 32-10 and 80-60 AACC (2003), while the fatty material content was evaluated according to the method of Schoch (1964). Amylose apparent and amylose contents were determined following the colorimetric procedure of Juliano (1971), and the calorimetric of Mestres et al., (1996). The ratio of Amylose/Amylopectin (A/Ap) was calculated from the equation:

$$Ratio\,\frac{amylose}{amylopectin}(A/P) = \frac{\text{amylose content (\%)}}{100 - \text{amylose contents (\%)}}$$

Purity was calculated from the difference between 100 and percent of moisture, crude protein, fatty material and ash content following the equation: % purity = 100 - [% crude protein + % fatty materials + crude fiber + % ash + % total sugars].

The density was evaluated by the method of Schoch and Leach (1964), as modified by González and Pérez (2002): sample size (2-3 g), and temperature (25 °C), while the pH and titratable acidity were determined by the official methods 02-52 and 02-31 AACC (2003). *Color* of the native and modified starches, expressed as ΔE was obtained by a Colorimeter (Hunter Lab, Reston, D-25). The color parameters are expressed in the CIE color scale L* a* b* (Hunter Lab, 2001; Manresa and Vicente, 2007). They were used to calculate the ΔE values of sample, comparing it to native starch parameters. Delta E (ΔE) was calculated from the equation:

$$\Delta E = \sqrt{\sqrt{(\Delta L^*)^2} + (\Delta a^*)^2 + (\Delta b^*)^2}$$

And also to calculate the white index (WI) of sample from the equation:

$$WI = 100 - \sqrt{(100 - L^*)^2 + a^{*2} + b^{*2}}$$

Syneresis was determined by visual observation of presence of water exudates from the gel (7%) after 24 hours at 4±1 °C.

Gel consistency was determined following the procedure of Cagampang et al., 1973.

Water absorption, solubility and swelling power: The water absorption, solubility and swelling power were evaluated according to procedure described by González and Pérez (2002).

Paste Clarity: Clarity of starch pastes were determined 25 °C, following the methodology described by Lim and Seib (1993). Near neutral pH aqueous slurry of starch (1%) is heated in a boiling water bath for 30 min., with intermittent shaking. Then the suspension is cooled for 1 hour at 25 ° C. The value of transmittance was read at 650nm in a Genesys 5 Spectrophotometer (Spectronic Instruments, Inc., Rochester, NY).

Damage starch: The percent of damage starch was determined according methodology described by Williams and Fegol, (1969), on samples of native starch and modified phosphate / acetylated.

Microstructure

Microstructure was analyzed by using different technique: Scanning Electron Microscopy (SEM) and Polarized Light Microscopy (PLM), size and size distribution.

Scanning Electron Microscopy (SEM): Each starch sample was spread on a piece of adhesive tape, double-sided, previously attached to the metallic holder. 2 drops of silver paint on one or more of the edges of the tape were added. These specimens were introduced for 25 min at 5 mA current in the gold coater (Ion Sputter, Eiko Engineering, IB2 Mod). Finally, the sample holder was placed in the Hitachi scanning electron microscope S-500; with an operating voltage of 20 kV, objective lens aperture 1 mm (No. 2), the condenser lens 540 and working distance of 15 mm. Photomicrographs were taken with a Mamiya (4 x 5) camera and a film Agfa Pan 100 Professional was used González-Parada and Mosqueda, 1996).

Polarized Light Microscopy (PLM): To assess the shape of the granules and the Maltese cross, a standard optical microscope Nikon Optihot 2 built brand Nikon FX photographic device was used- 35D and polarized light filter according to methodology described by Pérez et al., (2005).

Granule Size: Starch granule diameter was estimated by averaging the largest dimension of 30 random starch types on several field of SEM. Size categories were established by grouping granular diameters based on those sizes frequently found in the selected regions.

Granular size distribution: The granule size distribution of starch particles was performed at room temperature by laser diffraction using a Malvern Mastersizer 3000. A spatula tip of each starch sample was suspended in tap water and the suspension was directly fed into the granule size measuring cell at room temperature. The suspension at a standardized 2% laser obscuration was submitted to 10s sonication prior to laser diffraction measurement. Based on triplicate measurements, the granule size distribution was estimated using the Fraunhofer approximation considering opaque particles.

Pasting profile: The pasting profile were determined with the Amilógrafo Brabender (Brabender OHG, Duisburg, Typ. 800221, N° 930139) in an aqueous solution of 7% of concentration, following the method described in AACC (2003) N° 61-01. The breakdown, setback, and consistency indices were calculated from the corresponding plots. Four parameters were measured: pasting temperature (PT), peak viscosity (PV), hot paste viscosity at the end of the plateau at 90 °C (HPV) and the cool paste viscosity (CPV) at 50 °C. Three additional parameters were calculated: breakdown (BD) estimated as PV - HPV, setback (SB) estimated as CPV - PV, and consistency (CS), estimated as CPV - HPV (Rasper, 1980; Pérez et al., 2011).

X-Ray Diffraction Pattern: The spectra of X-ray diffraction of the experimental starches starches also were obtained from a diffractometer Bruker AXS D8 Advance model, equipped with a data analysis program. The operating conditions were as follows: Cu-Ka monochromatic radiation (λ = 1.542 Å); voltage: 40 kV, current: 40 mA, running speed: 0.05 ° 2θ/10s, 2nd angles were scanned at 40 ° and the rotational speed of the sample was 30 rpm. *The crystalline areas and the amorphous* fraction were quantified by fitting the diffraction patterns for win32 v4 PeakFit program (Jandel Scientific, USA). The relative degree of crystallinity was determined by the ratio of the total area of the crystalline fraction (peak) and the total area of the crystalline fraction of the area over the amorphous fraction, according to the following equation:

$$Relative\ crystallinity\ (\%) = \frac{Crystalline\ fraction\ area}{\text{Total area}}$$

X-ray diffraction patterns of the commercial starches were obtained with a X-ray diffractometer (Phillips) under the following conditions: monochromatic radiation of cobalt, voltage at 31kV, current at 25mA, time constant of 4 sec. and chart speed at 1 cm/min. Diffractograms were registered at $2\theta = 4–26\ ^{o}$ al a scan rate of 1°/min.

Data statistical analysis: The statistical package Statgraphics v. 6.0 (1992, Graphic Software Systems Inc., USA), and Anova-one way were used to determine differences among samples ($p < 0.05$), When there was differences, it was performed a multiple range test of Duncan.

Results and Discussion

Starches Isolated and Purified at the Laboratory from Cassava Clones of the Venezuelan Germplasm Bank

In Table 1, is showing the yield values of the starches isolated and purified from the edible portion of the roots. Starch in the edible portion of cassava roots, in wet weight basis, varied from 15 to 34% (INN, 1999, Tonukari, 2004, Fakir et al., 2012). The yields in the study varied from 14.08-16.64% and are agreed with the result of Fakir et al., 2012. The average starch yield for the 4 clones was not different ($p < 0.05$), but the trend was, from lowest to highest: 2192-2306-2681-2593 clone.

Table 2, summarize the proximate composition, percent of purity, amylose content and ratio /AP. As can see in Table, the starches have high purity over 99 %, with lightly differences only in ash and crude protein contents that can be by effect of unequal treatment of purification. The relation of amylose to amylopectin clearly is pointed out that these starches tend to be waxy one. The amylose content of these starches was determined by DSC; which give most accurate values. The moisture contents are in the range for a shell life stable (Radley, 1976). Moorthy (2002) indicates that the moisture content of starch varies from 6-16% depending on the drying process; highest content may lead to damage and subsequent microbial quality deterioration. Some authors reported protein content in cassava starches lower than those obtained in this investigation 0.14 % (Lorenz, 1982), 0.09% (Numfor and Noibi, 1995), 0.03% (Kaneda et al., 1996), 0.1% (Becker et al., 2001), 0.20% (Peroni et al., 2006), 1% (Zaidul et al., 2007).

Have also been reported protein contents above of those obtained in this research, such as those obtained by Matos-Segura and Pérez -Sira (1996) 0.75% and Alvis et al., (2008) between 0.60% and 0.61%. With regard to the fatty material content, the values were between 0.08% and 0.15%. On average no significant differences (p <0.05) among clone were observed. The phosphorus contents were 0.01% and 0.02%.

Table 1. Yield values of the starches isolated from the edible portion of the cassava roots

Sample / Clone	Yield (%)
2192	14.36±1.9[a]
2681	16.94±1.9[a]
2306	16.51±1.2[a]
2593	19.37±2.9[a]

Values are expressed as mean ± standard deviation. Values with same letters in a row have not are statistically difference ($\alpha \leq 0.05$).

Table 2. Proximate and chemical composition and purity of native and physical modified cassava starch

Parameters	2192	2681	2306	2593
Moisture (%)	8.73±0.43a	9.16±0.15a	8.99±0.04a	9.40±0.02a
Crude Protein (%)*	0.35±0.01a	0.31±0.04a	0.35±0.02a	0.25±0.01b
Fatty material (%)	0.12±0.07a	0.12±0.04a	0.11±0.04a	0.13±0.03a
Ash (%)	0.13±0.02a	0.17±0.01b	0.13±0.01a	0.13±0.01a
Phosphorous (%)	0.02±0.00a	0.02±0.00a	0.01±0.00a	0.01±0.00a
Purity (%)	99.40	99.40	99.41	99.49
Amylose (%)	17.26±0.50a	17.22±0.78a	17.99±0.47a	18.38±0.92a
Ratio Amylose/ Amylopectin	0.21	0.21	0.22	0.23

*: Dry basis **: Factor 6.25. Values are expressed as mean ± standard deviation. Values with same letters in a row have not are statistically difference ($\alpha \leq 0.05$).

On average no significant differences (p <0.05) were observed in the content of phosphorus with respect to the clones. Kaneda et al., (1996) reported that cassava starch containing 0.01% phosphorus, less than potato (0.09%), banana (0.03%), sweet potato (0.02%) and sago (0.02%). Gunaratne and Hoover (2002) reported 0.01%, Peroni et al. (2006) 0.007%, lower than other roots and tubers: 0.018% handsome, sweet potato 0.014%, 0.022% yam, sago 0.031% and jengibre 0.007%. As seen the results of the amylose contents here reported are agreed with those that have been reported in literature, which are ranging from 10.9 to 44.3% (Dakubu and Bruce-Smith, 1979; Pérez, 2000; Wuttisela et al., 2008; Mejía-Agüero et al., 2012).

Differences found in the amylose contents of the starch can be due to the analytical method used, age and/or plant variety (Peroni et al., 2006). Jane et al., 1999, (cited by Peroni et al., 2006) also have suggested that the differences may be attributed to the chain length of amylopectin, as longest are, they can form helical complex with iodine, given an overestimate of amylopectin content. Chen et al. (1998, cited by Kwaśniewska et al., 2008) have pointed out that the content of amylose and amylopectin in starch granules are genetically determined, but also, to some extent, they are affected by the environmental conditions, by regulating the activity enzymes involved in starch biosynthesis. One of the properties by definition of the starch is its white color "*Isolated starch is typically a dry, soft, white powder*". The white color is an index of quality and purity. In Table 3, can be seen the average of white index (WI). WI values are ranging from 97.06 and 97.85, corresponding to the first value to the darkest starch and the second to lightest ($p < 0.05$). However they are at all quite white, corroborating its high purity (Table 2). pH values also are among the quality parameters, due that the viscosity profile can vary as a function of the pH.

Table 3. pH and white index of the starches

Sample / Clone	White Index (WI)	pH
2192	97.06 ± 0.01^{a}	6.37 ± 0.03^{a}
2681	97.80 ± 0.01^{b}	6.51 ± 0.03^{a}
2306	97.85 ± 0.01^{b}	6.55 ± 0.02^{a}
2593	97.53 ± 0.01^{b}	6.80 ± 0.04^{b}

Values are expressed as mean ± standard deviation. Values with same letters in a row (a,b), have not are statistically difference ($\alpha \leq 0.05$).

In general pH extremes tend to have a negative impact on viscosity by hydrolyzing bonds and disrupting the molecular integrity if the granule (Thomas and Atwell, 1999). pH in the starches here studied are quite close to neutrality. Figure 1, is showing the water absorption, percent of solubility and swelling power of the native starches of experimental crops of cassava. These parameters are defining the functional behavior of the starch and are quite important in different steps of the processing.

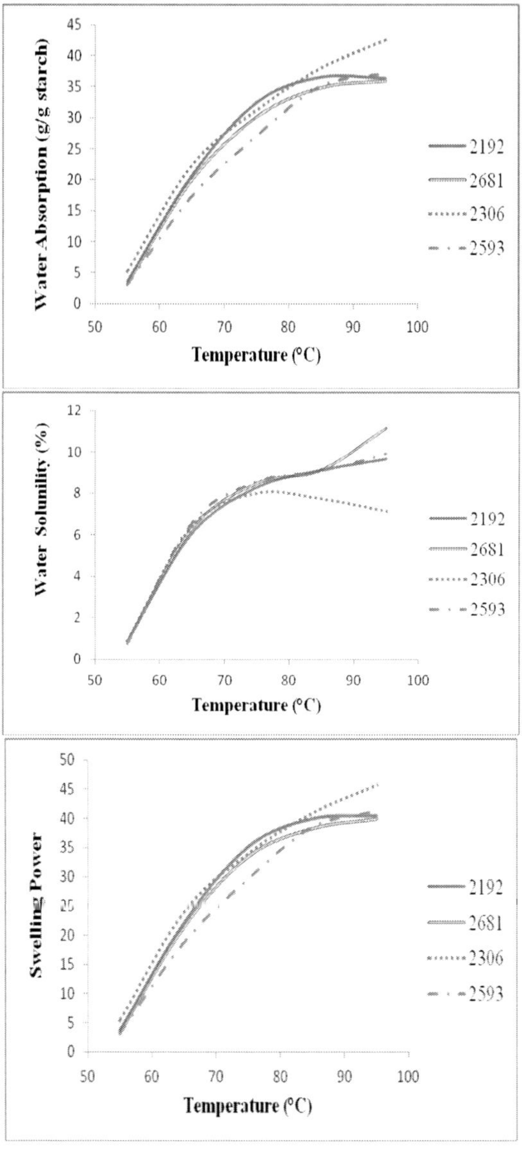

Figure 1. Water absorption, solubility and swelling power of native cassava starches.

As is showing the Figure 1, there are not major differences among the clones. However, the tiny differences observed among the clones could predict some differences of their structural and crystalline conformation. For examples, water penetration at the amorphous or crystalline lamellas or the force to permit the swelling.

The clarity of the paste is a property that depends on the amount of pigments, dirt and protein in the starch. It interacts with the degree of compaction of the polymers, especially amylose, during cooling. Table 4 summarizes the values obtained in this study for the starches.

Table 4. Paste clarity of the starches

Clones	Paste clarity
2192	59.17 ± 1.27^{ab}
2681	60.40 ± 1.23^{b}
2306	65.28 ± 0.22^{c}
2593	55.84 ± 0.79^{a}

Values are expressed as mean ± standard deviation. Values with same letters in a row (a,b), have not are statistically difference ($\alpha \leq 0.05$).

According to Aristizabal and Sanchez (2007), starch pastes with transmittance less than 40% are consider opaque or turbid, lie to, those with greater than 40% transmittance are considered clear or transparent. According to these authors, the pasta formed by the starches here studied is transparent, because their transmittance values are higher than 50%.

The scanning electron microscope (SEM) allows the study of the shape and size of the starch granules, as its size affects gelling properties, enzymatic susceptibility, crystallinity, swelling and solubility. The size of the granules is also a feature of importance in the extraction of the starch industry, and which will define the mesh size of the sieves for extraction and purification. As shown in Figure 2, the granules of the four clones are quite similar; with large egg truncate-shaped blunt ended, and some have a rounded shape, with smooth surface and without holes. The sizes are varying from 3 20 μm measured from the microscope camps. Similar results are cited by (Ciacco and D'Appolonia, 1976; Gunaratne and Hoover, 2002; Matos and Perez, 2003; González and Pérez, 2003; Ceballos et al., 2007). Ceballos et al., (2007) have also reported that cassava starch granules both "normal" and "waxy "have the same morphology. On the other hand, Ao and Jane, 2007, have pointed out that the rounded shape of the granules may result from a higher proportion of short chains in amylopectin.

Figure 2. Electron micrographs of experimental starch granules.

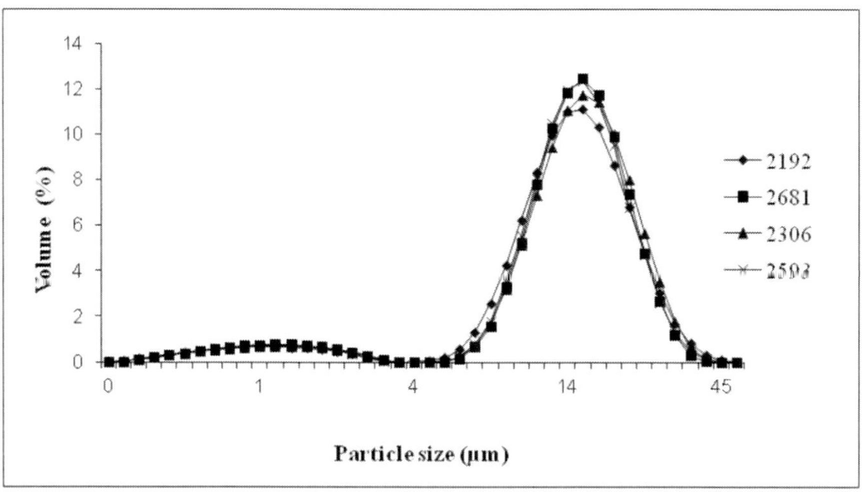

Figure 3. Granular size distribution measured by laser technique of the experimental starches.

Table 5. Gelatinization profile of the experimental starches

Parameters/BU	2192	2681	2306	2593
Breakdown	165.0	153.8	175.2	109.4
Setback	20.6	- 42.6	30.2	56.0
Consistency	144.4	196.4	205.4	165.4

The size granular distribution is view in Figure 3. The population is varying from 5 to 40 μm. The discrepancies between SEM and Laser technique, could be in regards to the forms of the granules; due the laser technique is considering round particles and the cassava starch are differently shaped. Other effect could be the aggregation of some granules for effect opt the purification process.

There exist notably differences among the clones in all of the gelatinization profile (Table 5). Clone 2681 starch showed the lowest values of retrogradation, fragility and consistency. Cassava starch has poor paste stability during prolonged cooking.

As shown in the graphs of intensity vs. 2θ, peaks 5 to 15, 17, 18, 20 and 23 ° have been developed as means of the four samples (Figure 4). Similar values were also observed by Gunaratne and Hoover (2002), and Matos and Pérez (2003) in native cassava starch. According to the study of the diffractograms, cassava starch of the 4 clones studied, starch is a type A because it shows a doublet at 17 and 18 ° 2θ and a singlet at 24 ° 2θ. Kaneda et al., (1996) and Srichuwong et al., (2005) also found that tapioca starch has a pattern of X-ray diffraction of type A. Buléon et al., (1998) point out that the observed differences in the plant species are related to the L/S ratio. It means the relation of the internal chains with degree of polymerization of 45-55 (called L) and short internal and external chains polymerization degree of 14-18 (called S), expressed on a molar basis. The authors have estimated that when the ratio is 5 it is B-type amylopectin starches (such as potatoes), and 8-10 for amylopectin starch type A. These authors also note that if amylopectin have high amount of chains with degree of polymerization between 6-12, show X ray diffraction patterns of A-type, and if they have a low amount, the pattern is B. Then it can be expected that amylopectin, of the starches here studied, have high amount of chains with degree of polymerization between 6-12,

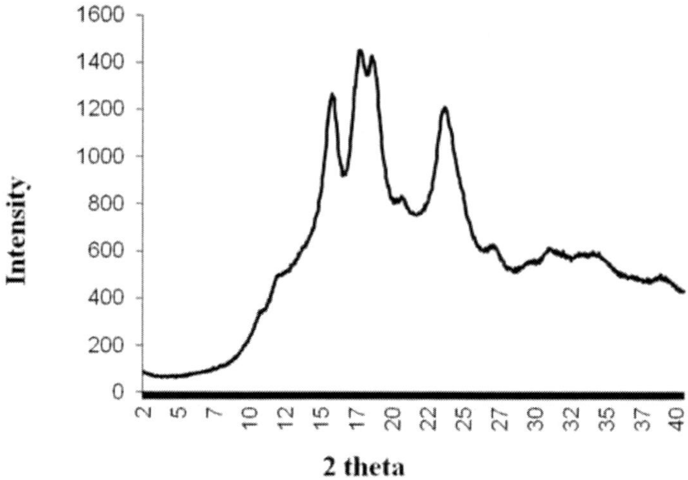

Figure 4. Spectra of X-ray diffraction of the experimental starches.

The relative crystallinity of native starches was measured by X-ray diffraction ranging from 28.89 to 33.55% (Table 5), Atichokudomchai et al., 2004 shown value of 39.53% for native cassava starch.

Table 5. Relative crystallinity of the experimental starches

Clones	Relative crystallinity (%)
2192	33.55
2681	29.68
2306	28.89
25.93	30.98

Table 6. Proximate and chemical composition and purity of native and physical modified cassava starch*

Parameters	Native	A	B	C	D	E	F	G
Moisture (%)	10.21	7.79	8.04	8.28	9.73	8.40	11.60	11.52
Crude Protein (%)	--	--	--	--	--	--	--	--
Fatty material (%)	0.12	0.08	0.08	0.00	0.01	0.07	0.12	0.12
Crude fiber (%)	0.28	0.29	0.29	0.15	0.29	0.18	NA	NA
Ash (%)	0.11	0.15	0.17	0.13	0.14	0.16	0.06	0.06
Total sugar (%)	0.02	0.06	ND	0.09	0.09	0.05	0.02	0.33
Reducing sugar (%)	0.02	0.03	ND	0.04	0.07	ND	0.02	0.14

Parameters	Native	A	B	C	D	E	F	G
Non-reducing sugar (%)	0.00	0.03	ND	0.05	0.02	0.05	ND	0.19
Purity (%)	99.47	99.42	99.46	99.63	99.57	99.44	99.80	99.49
Amylose (%)	41.59	45.20	42.89	41.87	41.36	49.52	NA	NA
Ratio Amylose/ Amylopectin (A/Ap)	0.71	0.83	0.75	0.72	0.71	0.48	NA	NA

*: Dry basis **: Factor 6,25
NR: Not Assayed.
ND: Not Detected.
A: Extrusion 1 screw, 90 rpm, 25% moisture content, 150°C.
B: Extrusion 2 screw, 90 rpm, 10,21% moisture content, 100°C.
C: Extrusion 2 screw, 90 rpm, 25% moisture content, 150°C.
D: Microwave: 50%, 6 min., 25% moisture content, 85°C.
E: Dehydration by drum dryer: 1, 2, 5, 8 rpm, 152°C.
F: Gamma Irradiation: 11.60% moisture content, 0.5Mrad.
G: Gamma Irradiation: 11.60% moisture content, 2.5 Mrad.

Commercial Native, Physical and Chemical Modified Starches

Table 6, is showing the effect of physical modification on the starch composition. As noted in the Table, the processes of extrusion, microwave and drum drying have decreased the contents of moisture, fatty material and crude fiber, as compared with the native starch, while gamma irradiation did not. A little increment of contents of ash and sugar were also observed by effect of extrusion, microwave and drum drying. The application of 2.5 Mrad has increased notably the contents of the total, reducing and non-reducing sugar in the starch, showing the rupture of the amylaceous components by effect of irradiation intensity. The amylose concentration was increased by the extrusion process of a screw, and drum drying, affecting for hence the ratio A/Ap. However, the percent of purity, at all, was not affected by the physical modification. The values of pH and titratable acidity were severely affected by the gamma irradiation (Table 7). It is corroborating the disruption effect of the irradiation that is most severe at 2.5 Mrad than 0.5 Mrad. The density values vary significantly only effect of dehydration with double drum drying and extrusion. The color was affected by the microwave and gamma irradiation. The gel consistency also was affected by all of the physical modification assayed. As the rheological parameters are shown in Tables 6 and 7, physical modification processes varied gel strength, texture, brittleness and settlement, the most relevant being the effect of gamma irradiation.

Table 7. Physical parameters of native and physical modified cassava starch

Parameters	Native	A	B	C	D	E	F	G
pH	5.4	5.6	5.3	5.7	5.4	5.8	4.2	3.5
Titratable acidity (NaOH meq)	0.018	0.013	0.013	0.013	0.017	0.010	0.95	2.50
Density (g/ml)	1.445	1.442	1.487	1.509	1.578	1.215	1.445	1.445
Color (ΔE)	--	6.31	1.17	1.83	8.70	4.17	25.25	25.95
Gel consistency (mm)	115	125	115	150	140	120	150	150
Syneresis (24 hours; 4±1 °C)	--	--	--	--	--	--	--	--

A: Extrusion 1 screw, 90 rpm, 25% moisture content, 150°C.
B: Extrusion 2 screw, 90 rpm, 10. 1% moisture content, 100°C.
C: Extrusion 2 screw, 90 rpm, 25% moisture content, 150°C.
D: Microwaves: 50%, 6 min, 25% moisture content, 85°C.
E: Dehydration by drum dryer: 1, 2, 5, 8 rpm, 152°C.
F: Gamma Irradiation: 11.60% moisture content, 0.5Mrad.
G: Gamma Irradiation: 11.60% moisture content, 2.5 Mrad.

Figure 5, reports the profile of the water absorption, water solubility and swelling power of the starches native and physically modified. As can see in Figure, except for the dehydration by drum dryer and gamma irradiation at 2.5Mrad, the water absorption remain quite similar to the native in the remaining starches. In regards to the dehydration by drum dryer it is quite obvious, due to the disruption of the granular structure by effect of the gelatinization, which will be observed later in the microphotography. An explanation for the diminution of water absorption by irradiation effect may be the transglucidation or formation of new chemical bonds in the amylaceous components. Figure 6, is comparing the microstructure of the native and modified starches. In this figure is observed that only the modification by dehydration with drum dryer that produce gelatinization of the starch, has disrupted the granular structure. Some tiny hole can be observed in some granule, as is pointed out by the arrow in Figure 6. The total disruption of the granular structure by the dehydration by drum dryer corroborated the pattern of water absorption, solubility and swelling power, shown by this starch.

Figure 7, is shown the pasting profile measured with the Amylograph Brabender of native, and physically modified cassava starches. All samples have changes in the viscosity peak, as compared to the native one. The extrusion has decreased the viscosity peak; but the effect most notably is by the moisture content/temperature, as is see in the B gelatinization profile, which is quite different from A and C, while A and C are quite similar in the pasting profile, despite they are different in the type of extrusion. The microwave irradiation effect is low, and its profile is similar to those of the native one. Dehydration with drum dryer has produced a starch with initial viscosity a 30°C, without viscosity peak and quite low consistency.

Despite that the two doses of gamma irradiation are showing, in the pasting profile, small viscosity peaks, have totally changed the pasting profile of the cassava starch. Table 8 summarizes the quantification parameters of the pasting profile. As can be seen, all of them have low retrogradation tendency and consistency. The gamma irradiation decreased notably the breakdown, setback and consistency as compared with the native starch.

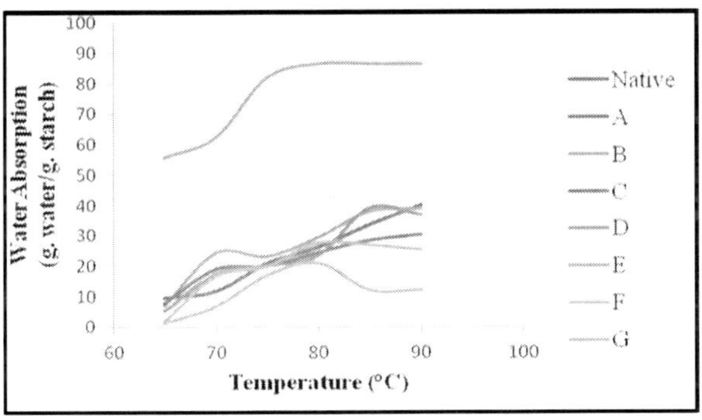

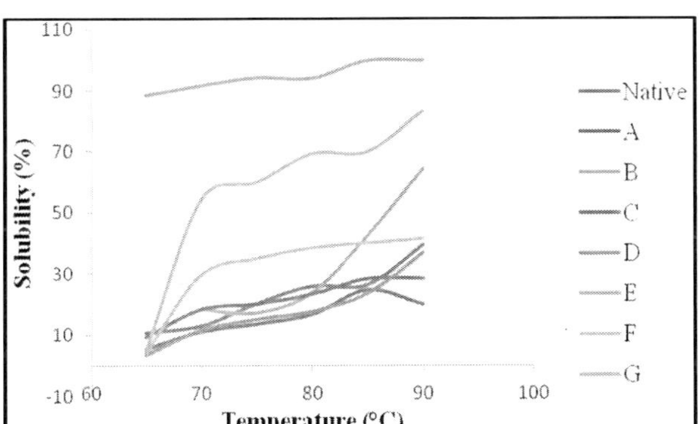

Figure 5. (Continued)

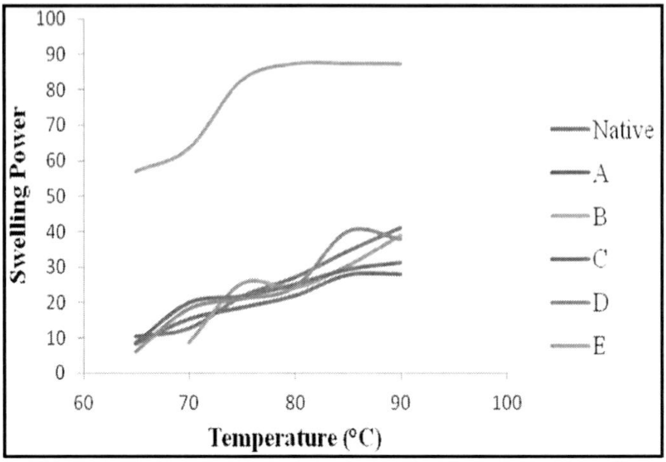

Figure 5. Water absorption, solubility and swelling power of native and physical modified cassava starches.

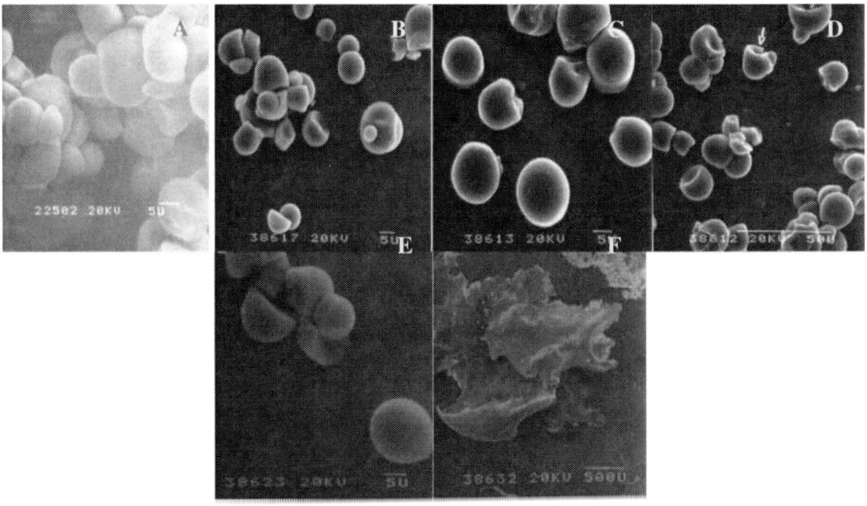

Figure 6. Microstructure measured by SEM of native and physical modified cassava starches. A: Native starch. B: Extrusion 1 screw, 90 rpm, 25% moisture content, 150°C. **C:** Extrusion 2 screw, 90 rpm, 10.21% moisture content, 100°C. **D:** Extrusion 2 screw, 90 rpm, 25% moisture content, 150°C. **E:** Microwaves: 50%, 6 min, 25% moisture content, 85°C. **F:** Dehydration by drum dryer: 1, 2, 5, 8 rpm, 152°C.

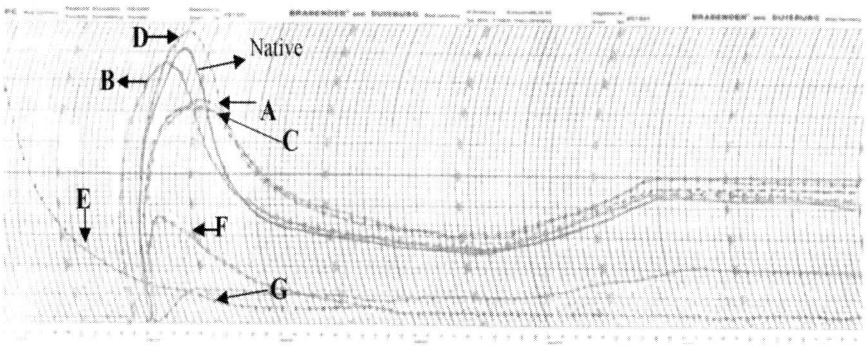

Figure 7. Pasting profile measured with the *Amylograph Brabender* of native and physical modified cassava starches. **A:** Extrusion 1 screw, 90 rpm, 25% moisture content, 150°C. **B:** Extrusion 2 screw, 90 rpm, 10.21% moisture content, 100°C. **C:** Extrusion 2 screw, 90 rpm, 25% moisture content, 150°C. **D:** Microwaves: 50%, 6 min, 25% moisture content, 85°C. **E:** Dehydration by drum dryer: 1, 2, 5, 8 rpm, 152°C. **F:** Gamma Irradiation: 11.60% moisture content, 0.5Mrad. **G:** Gamma Irradiation: 11.60% moisture content, 2.5 Mrad.

Table 8. Rheological parameters of native and physical modified cassava starch

Parameters/BU	Native	A	B	C	D	E	F	G
Breakdown	660	440	660	410	670	690	290	20
Setback	-500	-260	-440	-380	-500	-600	-290	-20
Consistency	160	180	160	30	170	170	0	0

BU: Brabender Units.

A: Extrusion 1 screw, 90 rpm, 25% moisture content, 150°C.

B: Extrusion 2 screw, 90 rpm, 10. 1% moisture content, 100°C.

C: Extrusion 2 screw, 90 rpm, 25% moisture content, 150°C.

D: Microwaves: 50%, 6 min, 25% moisture content, 85°C.

E: Dehydration by drum dryer: 1, 2, 5, 8 rpm, 152°C.

F: Gamma Irradiation: 11.60% moisture content, 0.5Mrad.

G: Gamma Irradiation: 11.60% moisture content, 2.5 Mrad.

Table 9. Proximate and chemical composition (dry basis) and purity of native and chemical modified cassava starch

Parameters	Native[3,4]	A[2]	B[3]	C[2]
Moisture (%)	8.2[3]-11.7[4]	6.6	11.55	12.33
Crude Protein (%)[2]	0.1[4]-0.7[3]	ND	0.06	0.5
Fatty material (%)	0.2	0.1	0.2	0.1
Crude fiber (%)	1.0	0.5	NA	0.5
Ash (%)	0.1	0.2	0.2	0.4
Purity (%)	99.0	99.3	99.0	98.6
Amylose (%)	18.88	ND	ND	NAP
Amylose Apparent (%)	34.13	35.1	NAP	36.6
Ratio Amylose/Amylopectin (A/Ap)	0.23	ND	ND	NAP
Ratio Amylose Apparent / Amylopectin (AA/Ap)	NAP	0.5	NAP	0.6

[1]Factor 6.25; [2]Matos y Pèrez, 1996, [3]Sívoli, 2009.

NA: Not Assayed.

ND: Not Detected.

NAP: Do not applied.

A: Crosslinked starch **B**: Crosslinked-Acetylated starch; C: Crosslinked commercial starch.

CHEMICAL MODIFICATIONS

Table 9 sums up the proximate composition, percent of purity, contents of amylose and apparent amylose, and ratio A/AP of native and modified starches. Starches have high purity over 99 %. However there exist tiny differences in the data of moisture and protein contents of the native and modified starches reported by the two authors. These little discrepancies can be attributed to the precision in the purification procedures. The moisture contents, of all starches, still are in the range for a shelf life stable (Radley, 1976). The fatty material contents of all starches averaged is 0.2% and are in the range reported in literature (Hoover, 2001, Atichokudomchai et al., 2001; Ellis, 1998; Jobling, 2004). The average ash content was 0.2%, it is low if is compared with the commercial one, and a value that corresponds to that reported by Gunaratne and Hoover (2002). The amylose contents were determined by calorimetric (DSC) and colorimetric procedures (Iodine chromogen- formation), the data in Table is denoting the differences among the two procedures, and it is remarking that one of the values is a true amylose content. It was determined by DSC; procedure that give most accurate values, and the other one (colorimetric determination) had overestimated this value. Despite these amylose results, the crosslinking of starch avoids its calorimetric determination in the modified starch, so they values could not be detected. As conclusion the experimental crosslinked starch is quite similar to those commercial.

Table 10 summarizes the phosphorous contents of the native, and degree of substitution (DG) of the phosphate-acetylated starches. A phosphorous content of cassava native starch was 0.06 % (w/w). The content of phosphorous in different starches is typically less than 0.08% (Lim et al., 1994). It have been reported, that 1 phosphate group typically exits per 200-400 glucose units in a starch molecule (Swinkels, 1985). The phosphate ester groups on starches contribute to its high clarity and viscosity, when it is cooked to a paste. Phosphatation of wheat and corn starches increased paste clarity and consistency, but potato starch remained superior in those properties in spite of the fourfold higher level of phosphate esters on the modified wheat and corn starches and their low level of cross-linking. The location of endogenous phosphate esters on potato starch apparently differs from that on chemically phosphated starches (Lim and Seib, 1993). Food and Drug Administration, HHS, 2013, in the § 172.892 establish as a limitation that residual phosphate and acetyl groups in food starch-modified not to exceed 0.4 % (calculated as phosphorus) and 2.5%, respectively**.** Table 10, shown that the phosphate-acetylated starches elaborated in this study are below these limitations

The size of the starch granules was not significantly different ($P<0.05$) across treatments. The granule sizes for the native and modified cassava starches were 3.0-16 μm and 6.0-18 μm respectively. Table 11, summarizes the size distribution of starch granules from different starches. The frequency of granules with diameters greater than 10 μm was greater for phosphorylated and acetylated starches. Although differences in sizes were not observed, chemical modifications can affect the pattern of granules size distribution.

Table 10. Phosphorus and acetyl contents and degrees of substitution (DS) for native and chemically modified cassava starches

Starch Type *	Phosphorus	DS	Acetyl (%)	DS
NS	0.06	0.003	0.0	0.0
AS	ND	ND	0.97	0.035
PS	0.31	0.013	ND	ND
PAS	0.18	0.006	1.66	0.064

*NS: native; PS: phosphorylated; AS: acetylated; PAS: phosphorylated-acetylated; [b] Not determinate.

Table 11. Granule size distribution of native and chemically modified cassava starches

Starch Type*	Diameter (μm)	Frequency (%)
NS	3-5.99	14
	6-9.99	41
	10-16	45
PS	6-9.99	28
	10-17	72
AS	6-9.99	42
	10-18	55
PAS	4-5.99	13
	6-9.99	42
	10-16	45

*NS: native; PS: phosphorylated; AS: acetylated; PAS: phosphorylated-acetylated; [b] Not determinate.

Table 12. Damage starch and Paste Clarity

Starch samples	Damage starch (%)	Paste Clarity (% T)
Native	2.05	52.58
Phosphate-acetylated	2.14	5.05

The clarity of starch reflected how light is transmitted through the paste. Compared with the native starch, the paste clarity of the dual-modified starch decreased significantly (Table 12). As previously reported cross-linking (Koo et al., 2010) decreased the paste clarity of starch, whereas acetylation significantly increased paste clarity of starches (Hung and Morita, 2005). Paste clarity is the result of rupture of swollen starch granules, and cross-linking improved the integrity of swollen granules, reducing paste clarity. Contrarily Xiao et al., 2012 report that compared with the native starch, the paste clarity of cross-linked starch decreased significantly, and the paste clarity of dual-modified starches significantly increased. That controversy could be attributed to the type reagents and/or starch type used. The clarity of the paste is affected by the remaining swollen granules, as a result of the chemical bonding by the crosslinking reflecting light through the paste.

Table 12 also has showing damage starch values of both the native and modified starch. Damage starch is an index of the quality of the starch and the effect of processing; and in this study of the modification. Damaged starch content in the native starch is about 2.05%; which is consistent with the value (2.33%) reported by Martinez-Bustos et al., 2007. The damaged starch content present in the modified starch is 2.14% showing that the starch did not undergo significant changes at the structural level during the modification.

X-Ray Diffraction Pattern: The X-ray diffraction patterns for the four types of cassava starches mentioned earlier were similar to each other. All yielded diffractograms with three major peaks at 2θ of 15, 17 and 23.5^{o} (4.21). Because modification not changes the X-ray patterns of the cassava starches, it may suggest that only the amorphous lamellae of the granules were reached by the chemicals.

Microscopic examination under polarized light (PLM): provides more detailed information in relation to the molecular organization of the starch granules. The birefringence naturally of the starch indicates the high degree of molecular orientation within the granules. In the MLP micrographic a well-defined *Maltese cross* is observed (Figure 8), as well as the edges of the granules, and small clusters. It can be seen the oval and round shapes-truncated granules with defined centric hilum; which indicates a high degree organization, due to the radial arrangement of the chains of amylose and amylopectin. González and Pérez, 2002, and Sodhi and Singh, 2005, reported that the surface of rice starch granules, there was rougher, and granules aggregates formed after acetylating than it counterpart native. This result is agreed with those observed in the micrographs of PLM in the modified samples; which shows the formation of aggregates in clusters; which are not observed in the native starches. Figure 9 shown the micrographic examination by SEM and it is corroborating the observation under polarized light; the view is showing the oval, and the round shapes-truncated granules (Figure 9A), and the agglomeration tendency (Figure 9B).

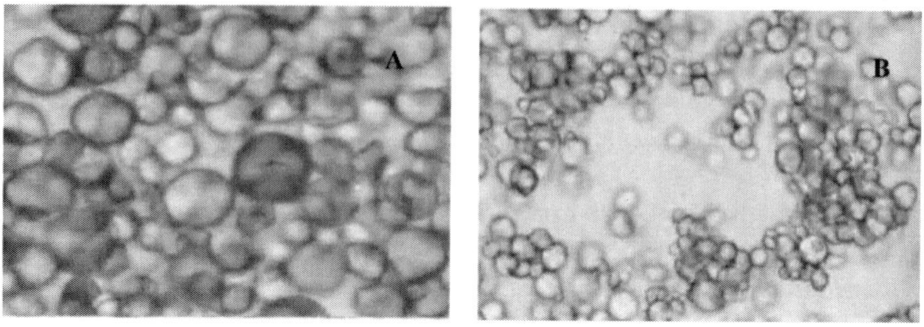

Figure 8. Micrographic examination under polarized light of the native (A) and chemical modified (B) starches.

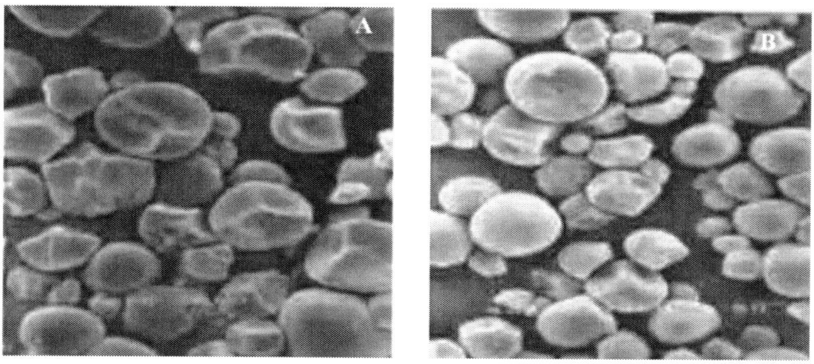

Figure 9. Micrographic examination SEM of the native (A) and chemical modified (B) starches (4 Kv).

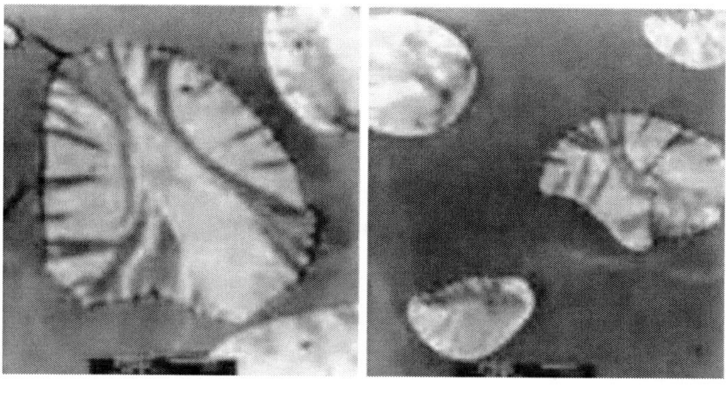

Figure 10. Micrographic examination TEM of the native (A) and chemical modified (B) starches.

Under the Transmission Electronic Microscopy (TEM) observation it is observed a number of internal channels in the granules (Figure 10) as well; numerous round structures adhered to the granule surface and located near the structures suggested as channels are observed. It could suggest that these channels could be the input of the material into the granule. It has been reported that the chemical treatment by acetylating alters slightly the granular morphology (Singh et al., 2004). These results are comparable with the results obtained in the present study, where only a slight increase in grain size was observed in the micrographs obtained using optical and electron microscopy. Same authors in a study conducted using corn and potato starches, and acetic anhydride as reagent, have reported that acetylation caused granular fusion, and it was more pronounced in potato starch, which are higher granule size.

Test at Pilot Scale (To Optimize the Industrial Process of Double Derivation of the Cross Linking/Acetylation)

After performing several running of the doble derivation (phosphate-acetylated) at pilot level, varying the parameters mentioned in materials and methods, it was reached a diminution of 5% of the sodium trimetaphosphate in the industrial processing. This amount is significant due to the large number of batches produced of this type of starch. Beside to save the reagent, it is reduce the pollution of the water dumped into the water treatment plant. The parameters established were at pH values between 10.8 and 11.0, and temperatures around 43 ° C, at which acceptable viscosity values were obtained in real time. (Sívoli, L., 2009).

Conclusion

The physical techniques of modification have affected notably the functional characteristics the starches, thus this one displays a greater versatility of use; not only in the food industry, but also in other industries.

Although the functional properties of the chemical modified starches were different as compared to native starches; their granules have same granular shapes, sizes and crystallographic patterns as its native counterpart. Indeed, when the starches, under the microscope of polarized light are observed, it can be distinguished the presence clearly of the Maltese cross, which is an indicative that the birefringence was not affected by the process of chemical modification. It is also observed, when using SEM and laser diffraction, in its granular structure, a smooth surface, without the pore presence and, a slight increase of size in the granular populations. Clarity of the paste had diminished by the reinforcing of the internal structure, caused by the chemical modification by cross linking. The patterns of X-rays diffraction of the starches chemically modified did not varied, when they are comparing to its native counterpart.

It was corroborated that temperature and pH are the main parameters driving the chemical modification by cross linking. Another parameters minor, but critical points are: agitation, way to add the reagents, characteristics of the raw material. All of them have to be controlled as parameters of quality control in the processing line. It was reached a diminution level of 5% in the trimetaphosphate main reagent in this kind of chemical modification. It is a representative amount of save that is incident in the diminution of cost and pollution.

REFERENCIAS

AACC, Cereal Laboratory Methods. (2003). The American Association of Cereal Chemists, Inc., St. Paul, MN USA. Methods: 02-31, 02-52, 08-12, 32-10, 46-13, 44-19, 80-60.

Alvis, A., Vélez, C., Villada, H. & Rada, M. (2008). Análisis físico-químico y morfológico de almidones de ñame, yuca y papa y determinación de la viscosidad de las pastas. *Información Tecnológica*, *19*(1), 19-28.

Ao, Z. & Jane, J. L. (2007). Characterization and modeling of the A- and B-granule starches of wheat, triticale, and barley. *Carbohydrate Polymers*, *67*, 46-55.

Aristizábal, J. & Sánchez, T. (2007). Guía técnica para producción y análisis de almidón de yuca. Boletín de servicios agrícolas de la FAO nº 163. Organización de las Naciones Unidas para la agricultura y la Alimentación, Roma.

Atichokudomchai, N., Shobsngob, S., Chinachoti,P. & Varavinit S. (2001). A study of some physicochemical properties of high crystalline tapioca starch. *Starch/Starke*, *53*, 577-581.

Atichokudomchai, N., Varavinit, S. & Chinachoti, P. (2004). A study of ordered structure in acid–modified tapioca starch by ^{13}C CP/MAS solid state NMR. *Carbohydrate Polymers*, *58*, 383-389.

Becker, A., Hill, S. & Mitchell, J. (2001). Relevance of amylose-lipid complexes to the behavior of thermally processed starches. *Starch/Stärke*, *53*, 121-130

Beleia, A., Butarelo, S. & Silva, R. (2006). Modeling of starch gelatinization during cooking of cassava (*Manihot esculenta Crantz*). *LWT-Food Science and Technology*, *39*, 399-404.

Buléon, A., Colonna, P., Planchot, V. & Ball S. (1998). Starch granules: structure and biosynthesis. *International Journal of Biological Macromolecules*, *23*, 85-112

Cagampang, G. B., Perez, C. M. & Juliano, B. O. (1973). A gel consistency for eating quality pf rice. *Journal of Science and Food Agriculture*, *24*(2), 1589-1594

Ceballos, H., Sánchez, T., Morante, N., Fregene, M., Dufour, D., Smith, A., Denyer, K., Pérez, J. C., Calle, F. & Mestres, Ch. (2007). Discovery of an Amylose – free Starch mutant in cassava (Manihot esculenta Crantz). *Journal of Agricultural and Food Chemistry*, *55*, 7469-7476.

Choi, S. & Kerr, W. (2002). ^{1}HNMR studies of molecular mobility in wheat starch. *Food research International*, *36*, 341-348.

Ciacco, C. F. & D'Appolonia, B. L. (1978). Baking studies with cassava and yam flour. I. Biochemical composition of cassava and yam flour. *Cereal Chemistry*, *55*, 402 – 411

Ciarfella, A. T. (2009). Evaluación de las raíces comestibles de cuatro variedades promisorias de yuca, caracterización del almidón y elaboración de casabe fortificado. Tesis de Doctorado, Universidad Central de Venezuela.

Colonna, P., Doublier, J. L., Melcion, J. P., de Monredon, F. & Mercier, C. (1984). Extrusion cooking and drum drying of wheat starch. I. Physical and macromolecular modifications. *Cereal Chemistry*, *61*, 538-543.

Dakubu, M. & Bruce-Smith, S. P. (1979). Amylose content of starch from different varieties of cassava (*Manihot esculenta* Crantz) in Ghana. *Ghana Journal of Agriculture Science*, *12*, 143-145

Doublier, J. L., Colonna, P. & Mercier, C. (1986). Extrusion cooking and drum drying of wheat starch II. Rheological characterization of starch pastes. *Cereal Chemistry*, *63*, 240-246.

Ellis, R. P., Cochrane, M. P., Dale, M. F. B., Duffus, C. M., Lynn, A., Morrison, I. M., Prentice, R. D. M., Swanston, J. S. & Tiller, S. A. (1998). Starch production and industrial uses. *Journal of the Science of Food and Agriculture*, *77*(3), 298-311.

Engelmann, G., Bonatz, E., Bechthold, I. & Rafler, G. (2001). Preparation of starches carbamates in homogenous phase using different mixing conditions. *Starch/Starke*, 53, 560-569.

Fakir, M. S. A., Jannat, M., Mostafa, M. G. & Seal1, H. (2012). Starch and flour extraction and nutrient composition of tuber in seven cassava accessions *Journal of the Bangladesh Agricultural University*, *10*(2), 217–222

FDA, U.S. Food and Drug Administration. (2013). U.S. HHS. Part 172-Food additives permitted for direct addition to food for human consumption Subpart I- Multipurpose Additives. Sec. 172.892. Food starch.

Fortuna, T., Januszewska, R., Juszczak, L., Kielski, A. & Palasinski, M. (2000). The influence of starch pore characteristics on pasting behaviour. *International Journal of Food Science and Technology*, *35*, 285-291.

González, Z. & Pérez E. (2002). Effect of acetylation on some properties of rice starch. *Starch/Starke*, *54*, 148-154.

González, Z. & Pérez, E. (2003). Evaluación fisicoquímica y funcional de almidones de yuca (*Manihot esculenta* Crantz) pregelatinizados y calentados con microondas. *Acta Científica Venezolana*, *54*, 127-137.

González-Parada, Z. & de Mosqueda, M. B. (1996). Estudio en el microscopio electrónico de barrido de harinas precocidas de maíz. *Tecnología Alimentos (México)*, *31*, 5-9

Gunaratne, A. & Hoover, R. (2002). Effect of heat-moisture treatment on the structure and physicochemical properties of tuber and root starches. *Carbohydrate Polymers*, *49*, 425-437

Hoover, R. (2001). Composition, molecular structure, and physicochemical properties of tuber and root starches: A review. *Carbohydrate Polymers*, *45*(3), 253–267.

Huber, K. C. & BeMiller, J. N. (2000). Channels of maize and sorghum starch granules. *Carbohydrate Polymers*, *41* (3), 269-276.

Hung, P. V. & Morita, N. (2005). Effects of Granule Sizes on Physicochemical Properties of Cross-linked and Acetylated Wheat Starches. *Starch/Stärke*, *57*, 413–420

Hunter Lab Manual. (2001). *Hunter associates laboratory universal software version* 3.8. ISO 9001 Certified. 55p.

INN (Instituto Nacional de Nutrición). (1999). Tabla de Composición de Alimentos. (9na ed.) Serie Cuadernos Azules. Caracas. Venezuela: Instituto Nacional de Nutrición.

Jobling, S. (2004). Improving starch for food and industrial applications. *Current Opinion in Plant Biology*, *7*(2), 210-218

Juliano, B. (1971). Simplified assay for milled rice amylose. *Cereal Science Today*, *16*, 334-360.

Jyothi, A., Moorthy, S. & Rajasekharan, K. (2006). Effect of Cross-linking with Epichlorohydrin on the Properties of Cassava (Manihot esculenta Crantz) *Starch/Stärke*, *58*, 292-299.

Kaneda, Y., Kitahara, K., Suganuma, T. & Nagahama T. (1996). Introduction of fatty acids into low-lipid starches and their nägeli amylodextrins. *Cereal Chemistry*, *73* (2), 244-248

Koo, S. H., Lee, K. Y. & Lee, H. G. (2010). Effect of cross-linking on the physicochemical and physiological properties of corn starch. *Food Hydrocolloids*, *24*, 619-625.

Kwaśniewska, I., Nebesny, E. & Rosicka, J. (2008). Characterization of amylose-lipid complexes derived from different wheat varieties and their susceptibility to enzymatic hydrolysis. *Food Science and Technology*, *14*, 29-37.

Lawal O. (2004). Succinyl and acetyl starch derivatives of a hybrid maize: physicochemical characteristics and retrogradation properties monitored by differential scanning calorimetry. *Carbohydrate Research*, *339*, 2673-2682.

Lewandowicz, G., Fornal, J. & Walkowski, A. (1997). Effect of microwave radiation on physico-chemical properties and structure of potato and tapioca starches. *Carbohydrate Polymers*, *34*, 213-230.

Lim, S. T., Kasemsuwan, T. & Jane, J. L. (1994). Characterizationof phosphorus in starch by ^{31}Pnuclear magnetic resonancespectroscopy. *CerealChemistry*, *71*, 488-493.

Lim, S. T. & Seib P. A. (1993). Location of phosphate esters in a wheat starch phosphate by ^{31}Pnuclear magnetic resonance spectroscopy. *Cereal Chemistry*, *70*, 145-152.

Liu, H. J., Ramsden, L. & Corke, H. (1999). Physical properties of cross-linked and acetylated normal and waxy rice starch. *Starch/Starke*, *51*, 249-252.

Lorenz, K. (1982). Cereal and root starch modification by heat-moisture treatment. I. physicochemical properties. *Starch/Stärke*, *34*, 50-54

Manresa, A. & Vicente, I. (2007). Medición del color. En: El color en la industria de alimentos. Ed. Universitaria, La Habana, Cuba. p. 64

Martínez-Bustos, F., López-Soto, M., San Martín-Martínez, E., Zazueta-Morales, J. J. & Velez-Medina, J. J. (2007). Effects of high energy milling on some functional properties of jicama starch (Pachyrrhizus erosus L.

Urban) and cassava starch (Manihot esculenta Crantz). *Journal of Food Engineering*, *78*, 1212-1220.

Matos, M. & Pérez, E. (2003). Characterization of native and modified cassava starches by scanning electron microscopy and X-ray diffraction techniques. *Cereal Foods World*, *48* (2), 78-8

Matos-Segura, M. E. & Pérez-Sira, E. (1996). Comparación de las propiedades del almidón de yuca (*Manihot esculenta* Crantz) fosfatado a escala de laboratorio con uno comercial. *Tecnología de Alimentos México*, (*6*), 23-27.

McKee, T. & McKee, J. (2003). Bioquímica. La base molecular de la vida. 3era. Edición. Mc. Graw-Hill Interamericana.

Mejía-Agüero, L. E., Galeno, F., Hernández-Hernández, O., Matehus, J. & Tovar, J. (2012). Starch determination, amylose content and susceptibility to *in vitro*amylolysis in flours from the roots of 25 cassava varieties. *Journal of the Science of Food and Agriculture*, *92*(3), 673–678.

Mestres, C., Matencio, F., Pons, B., Yadij, M. & Fliedel, G. (1996). A rapid method for the determination of amylose content by using differential scanning calorimetry. *Starch/Stärke*, *48*, 2-6.

Moorthy, D. (2002). Physicochemical and functional properties of tropical tuber starches: a review. *Starch/Stärke*, *54*, 559-552

Nabeshima, E. H. & Grossmann, M. V. E. (2001). Functional properties of pregelatinized and cross-linked cassava starch obtained by extrusion with sodium trimetaphosphate. *Carbohydrate Polymers*, *45*, 347-353.

Numfor, F. & Noubi, L. (1995). Effect of full-fat soya bean flour on the quality and acceptability of fermented cassava flour. *Food and Nutrition Bulletin*, *16* (3), 244.

Paschall, E. F. (1964). Phosphorylation with organic phosphate salts. *Methods in Carbohydrate Chemistry*, *4*, 214-296.

Pérez, E. (1996). Algunas experiencias sobre la modificación de almidones realizados en Venezuela. En: Memorias de Conferencias Internacional: Almidón, propiedades fisicoquímicas, funcionales y nutricionales. Ed. Escuela Politécnica Nacional de Ecuador. Quito, Ecuador. 129-148

Pérez, E. (2000). Determination of the correlation between amylose and phosphorus content and gelatinization profile of starches and flours obtained from edible tropical tubers using differential scanning calorimetry and atomic absorption spectroscopy. Thesis--PlanA (M.S.)--University of Wisconsin—Stout.

Pérez, E., Schultz, F. & Pacheco de, Delahaye E. (2005). Characterization of some properties of starches isolated from *Xanthosoma saggittifolium* (tannia) and *Colocassia esculenta* (taro). *Carbohydrate Polymers, 60* (2), 139-145.

Pérez, E., Gibert, O., Sabate, A., Jiménez, Y., Sánchez, T., Giraldo, A., Pontoire, B., Guilois, S., Lahon, M., Reynes, M. & Dufour, D. (2011). Physicochemical, functional and macromolecular properties of waxy yam starches discovered from "Mapuey" (*Dioscorea trifida*) genotypes in the Venezuelan Amazon. *Journal of Agricultural and Food Chemistry, 59*, 263–273.

Peroni, F., Rocha, T. & Franco, C. (2006). Some structural and physicochemical characteristics of tuber and root starches. *Food Science and Technology, 12*, 505-513

Radley, J. (1976). Examination and analysis of starch products. Applied Sciences Publishers LTD London.

Rasper, V. (1980). Theoretical aspects of amylographology In the *Amylograph Handbook*, the American Association of Cereal Chemists, Inc., St. Paul, 1-6.

Schoch, T. J. (1964). *Methods in Carbohydrate Chemistry*, Academic Press, Inc., New York, Vol. *IV*, 56-61, and 106-108.

Schoch, T. J. & Leach, H. W. (1964). Determination of absolute density. In *Methods in Carbohydrate Chemistry*, Academic Press, Inc., New York, Vol. *IV*, 101-103.

Singh, N., Chawla, D. & Singh, J. (2004). Influence of acetic anhydride on physicochemical, morphological and thermal properties of corn and potato starches. *Food Hydrocolloids, 17*, 63-72.

Sívoli, L. (2009). Cambios en la estructura granular, molecular y cristalina del almidón de yuca por efecto de la modificación química (fosfatación – acetilación). Tesis Doctoral en Ciencia y Tecnología de Alimentos, Universidad Central de Venezuela.

Slattery, C. J., Kavakii, I. H. & Okiro, T. W. (2000). Engineering starch for increases quantity and quality. *Trends in Plant Science, 5*(7), 291-298.

Sodhi, N. S. & Singh N. (2005). Characteristics of acetylated starches prepared using starches separated from different rice cultivars. *Journal of Food Engineering, 70*, 117-127.

Srichuwong, S., Candra, T., Mishima, T., Isono, N. & Hisamatsu M. (2005). Starches from different botanical sources. I: Contribution of amylopectin fine structure to thermal properties and enzyme digestibility. *Carbohydrate polymers, 60*, 529-538.

Swinkels, J. (1985). Sources of starch its chemistry and physics. In *G.M.A. Van Beymun* (ed.), Starch Conversion Technology. A. Rods. Pub. New York, Marcel Dekker.

Thomas, D. J. & Atwell, W. A. (1999). Starch structure In: Starches practical guide for the food industry. Eagan Press guide for the food industry. Eagan Press Handbook series. St. Paul Mn USA 1-12

Tonukari, N. J. (2004) Cassava and the future of starch. *Electronic Journal of Biotechnology*, *7*(1), 5-7 In: http://www.ejbiotechnology.info/

Whistler, R., BeMiller, J. & Paschall, E. (1984). Starch: *Chemistry and Technology* (2nd. Ed.) Academic, Orlando, 313-314.

Williams, P. & Fegol, K. (1969). Colorimetric determination of damaged starch in flour. *Cereal Chemistry*, *46*, 56-62.

Wuttisela, K., Shobsngob, S., Triampo, W. & Triampo, D. (2008). Amylose/amylopectin simple determination in acid hydrolyzed tapioca starch. *Journal of the Chilean Chemical Society*, *53* (3), 1565-1567.

Wurzburg, O. B. (1986). Crosslinking starches In: Modified Starches: Properties and Uses (edited by OB Wurzburg) Boca Ratón CRC Press, 41-53.

Xiao, H-X., Lin, Q-L., Liu, G-Q. & Yu, F-X. (2012). A Comparative Study of the Characteristics of Cross-Linked, Oxidized and Dual-Modified Rice Starches. *Molecules*, *17*, 10946-10957.

Zaidul, I., Nik, N., Mohd, A., Yamauchi, H. & Noda T. (2007). Analysis of mixtures of wheat flour and potato, sweet potato, yam and cassava starches. *Carbohydrate polymers*, *69*, 784-791.

In: Cassava
Editor: Francis P. Molinari

ISBN: 978-1-63321-031-8

Chapter 4

CASSAVA FLOUR AND STARCH AS DIFFERENTIATED INGREDIENTS FOR GLUTEN FREE PRODUCTS

Cecilia Dini[1], María Cecilia Doporto[1], Sonia Zulma Viña[1,2] and María Alejandra García[1,*]
[1]Center of Research and Development in Food Cryotechnology (CIDCA-CONICET-UNLP), CCT La Plata, Argentina
[2]College of Agricultural and Forest Sciences, National University of La Plata, Argentina

ABSTRACT

Food industry is interested in the study of sub-utilized plant species as a contribution to consumers' diet diversification and the incorporation of ingredients with specific technological properties.

In Argentina, more than 180,000 tonnes per year of cassava roots are harvested and this production supplies the domestic consumption and part of the local starch market. The Food and Agriculture Organization of the United Nations (FAO) has pointed out that many developing countries could strengthen their rural economy and increase farmers' income by converting relative low cost cassava feedstock in high-value starch. Compared to those obtained from other sources, cassava starch has more

[*] E-mail: magarcia@quimica.unlp.edu.ar.

clear and viscous pastes and is stable in acidic food products. It can be used in pharmaceuticals and natural thermoplastic products.

In this chapter, data referred to cassava production, consumption and processing as well as some results obtained from the technological and nutritional characterization of cassava roots and their derived starch and flour are presented. Cassava can be considered a very good source of starch and might be used as a functional ingredient, particularly for the formulation of gluten free products.

CASSAVA PRODUCTION

Cassava Production Characteristics

Cassava (*Manihot esculenta* Crantz) is grown in over 90 countries mainly in Africa, Asia, and Latin America, being a staple food for half a billion people. After rice and maize grains, cassava roots and their derivatives are the third most important source of calories in tropical and subtropical areas of the world. Table 1 shows the top producing countries in Africa, Asia, and America, in terms of harvested area and production in 2012.

Nigerian cassava production is by far the largest in the world, being more than twice higher than the production of Indonesia, Brazil, or Thailand (Table 1). Cassava production in other African countries, such as the Democratic Republic of the Congo, Ghana, Angola, Mozambique, the United Republic of Tanzania, Uganda, Cameroon, Malawi, and Madagascar seems undersized in comparison to the substantial amount produced by Nigeria. However, most cassava production in this country (about 90%) is intended for domestic food use and produced by smallholder farmers. Production yields are still low (14,026 kg/ha) [1].

In sub-Saharan Africa and many other parts of the world, typical cassava crop conditions correspond mostly to subsistence systems, often developed on marginal terrains. In those systems, roots are grown for food by small-scale, low-income farmers who sell the surplus production. In this context, cassava is fundamental for both earning generation and food security.

Gender contribution to cassava production in rural areas is relevant. In general, women represent an active labour force and they play a significant role although they rarely own the means of production. In this sense, there are some experiences oriented to capitalize on these resources. In order to meet the increased demand for convenient and affordable traditional food as well as to promote local markets, The Rural Women Foundation (RWF, Nigeria,

Africa) released a Project (Cassava Factory) that was aimed to enhance the capacity of the women small holder farmers and vulnerable groups to face the opportunities in the domestic market [2]. It has been pointed out that women are the primary and most clued-up producers. However, their access to enhanced processing technology is often slanted. According to the information provided by RWF, the cassava factory located in Ukwa West Local Government Area in Abia State targeted the development of the rural women group (Asa North Women Cooperative Society) with the scope of improving their incomes and well being.

In addition to produce food, administer natural resources and sustain economically their families, women are generally the key caretakers of household food safety. Also, they are mainly responsible for preparing meals, rationing food and take care of their children's health.

Table 1. Cassava harvested area (ha) and production (tonnes) for the first main cassava producer countries in Africa, Asia, and America in 2012

Continent	Country	Harvested area (ha)	Production (tonnes)
Africa	Nigeria	3,850,000	54,000,000
	Democratic Republic of The Congo	2,200,000	16,000,000
	Ghana	868,550	14,547,279
	Angola	1,062,865	10,636,400
	Mozambique	762,598	10,051,364
	United Republic of Tanzania	954,509	5,462,454
Asia	Indonesia	1,119,784	23,922,075
	Thailand	1,250,000	22,500,000
	Viet Nam	550,810	9,745,546
	India	223,000	8,120,000
	Cambodia	337,065	7,613,697
	China	280,000	4,560,000
America	Brazil	1,692,986	23,044,557
	Paraguay	183,000	2,560,000
	Colombia	211,714	2,274,358
	Peru	92,425	1,119,560
	Haiti	156,000	660,000
	Costa Rica	38,000	560,000

Source: FAOSTAT | © FAO Statistics Division 2014.

Thus, an important issue is that training women in order to improve farm yields will have positive impact in family welfare and nutrition. In addition to generate income, concerted actions should produce foodstuffs contributing to a balanced diet and to eradicate hunger and malnutrition.

With some different productive characteristics, large areas of tropical and subtropical Asia and the southern sub-tropics of South America are going over a developing process. In this context, cassava production chain is changing from a traditional food crop to an export oriented industry, particularly in Vietnam and Thailand [3]. Cassava roots are also an important source of starch in Southern Brazil, Paraguay, and the Northeast of Argentina.

Previously, cassava expansion was normally based on augmented cropping area. More recent changes in Southern China, Thailand, and Cambodia are based on both increased planted area and intensification of the crop, meanwhile in Southern India, Indonesia, and Vietnam more intensive production systems became prevalent [3].

Added to root production, cassava is also a source of animal feed, fiber for paper and textile manufacturing, and starch for food and pharmaceutical industries. Consequently, there is a changing approach and cassava potential to meet the needs of a rapidly-developing world facing the challenges of climate change is being revaluated. The crop could be adapted to an intensification process.

The Food and Agriculture Organization of the United Nations (FAO) has pointed out that many developing countries could strengthen their rural economy and increase farmers' income by converting relative low cost cassava feedstock in high-value starch. Compared to those obtained from other sources, cassava starch has more clear and viscous pastes and it is stable in acidic food products. It can be used in pharmaceuticals and natural thermoplastic products.

The strategy for cassava development, sponsored by the International Fund for Agricultural Development and FAO, gives value to the roots processing with a focus on the market-driven chain. While cassava starch global demand grew since 30 years ago, only Thailand evolved from the production of basic food to the processing industry. World-wide studies show that tropical countries imported each year corn starch and derivatives for more than 80 million U$S that could be replaced with locally obtained starch or cassava flour [4].

According to FAO data, cassava production in 2012 achieved 187,000 tonnes in Argentina [1] (Figure 1). This production supplies the domestic consumption and 75-80 % of the local starch market.

a

b

Figure 1. a- Cassava crop in Misiones, Argentina; b- Cassava roots. This work is a derivative of: by La.Catholique (contact).

Prospective Nutritional Topics

Agricultural and livestock products are the primary source of nutrients, so they must contain adequate quantities of all nutrients in order to support healthy grow and development of individuals. Unfortunately, many agricultural systems in developing countries do not fit with this requisite.

According to Welch and Graham [5], during the end of 20th century scientists at several Consultative Group on International Agricultural Research (CGIAR) Centers have been collecting data on the potential for breeding to significantly increase the levels of bioavailable iron, zinc, and provitamin A carotenoids in edible portions of cereals, beans, and cassava.

Referring to cassava, the variation in β-carotene concentration in roots from 630 genotypes collected at International Center for Tropical Agriculture (CIAT) was reported by Iglesias et al., [6]. Authors indicated β-carotene levels in fresh roots ranging from 0.1-2.4 mg/100 g [7]. The concentration of β-carotene appears to be determined by the genes that control the transport of precursors from the shoot to the roots and the biochemical processes affecting its further accumulation. The intensity of the orange color seemed to be a feasible visual criterion. Iglesias et al., [6] pointed out that it would be possible to obtain cassava roots with β-carotene concentrations high enough to meet the diet daily requirements on the basis of the available genetic variability within cassava germplasm. In this sense, cassava lines collected from the Amazonian region of Brazil and Colombia showed the highest levels of β-carotene.

Yeoh and Truong [8] pointed out that for people who consume cassava as a staple food its low protein content (PC) is critical, especially if their diets do not include protein from other sources. For 15 cassava varieties, PC ranged from 5 to 19 g/kg dry matter. Moreover, PC might vary according to root developmental stage and environmental conditions.

Later, Ceballos et al., [9] analysed cassava root total crude PC from 149 accessions belonging to the germplasm collection at CIAT and they found large differences in PC (ranging from 0.95% to 6.42%) in the analyzed samples. The authors concluded that the results were certainly auspicious, especially considering that the world cassava germplasm collection contains more than 6,000 accessions. Therefore, the possibilities of further increasing the natural range of variation for PC through traditional recurrent selection methods are also very promising [9].

Cassava Roots and Derived Products

Typical Meals and Nutritional Issues

In countries where cassava is a common crop many autochthonous meals based on this root have been developed, some of which are nowadays considered classic recipes of each regional culinary culture.

Figure 2. Cassava fried sticks. This work is a derivative of: (cc) by fczuardi (contact).

Some cassava meals are prepared from fresh roots: roasted, as fried sticks (just as French-fried potatoes; Figure 2), "Cassava Mofongo" (mashed fried cassava with spices, garlic and bacon), "creamed cassava" (mashed yucca, sometimes flavoured with garlic), Cuban "yuca con mojo" (cassava with garlic sauce) which is usually served as a side dish, and desserts such as cassava cake and Bibingka (Cassava Custard), classic Filipino recipes made from grated cassava, among others.

In West Africa, cassava is processed predominantly by grating the peeled roots and allowing fermentation for about 3 days, followed by expression of the cyanogenic liquid and heating with stirring to give a granular, roasted product called "gari" [10]. Cassava leaves stew is also widely consumed in west and central Africa.

Cassava processing is indispensable since it is one of the plants most commonly associated to cyanide poisoning.

Cyanide (hydrogen cyanide, HCN) is usually present in the plant kingdom as cyanogenic glycosides (or cyanogens), which are glycosides of sugars and a cyanide containing aglycon [11]. Disruption of the plant cells allows cyanogens to enter in contact to enzyme glycosidase, present in a different compartmentalization of the plant cell, which catalyses their breakdown to sugars and cyanohydrins. Cyanohydrins are decomposed above pH 5 leading to the production of hydrogen cyanide (HCN) and a ketone or aldehyde [10].

Linamarin and lotaustralin (methyl linamarin) are cassava predominant cyanogenic glycosides, both disrupted by the enzyme linamarase [10; 12].

Cyanide level widely depends on cassava variety, the plant age, since young plants contain higher amounts of cyanides than the mature ones [11], and the agroclimatic conditions of cultivation [13]. Reported values of hydrogen cyanide in cassava roots vary from 15-400 mg/kg of fresh weight [14].

Cyanogenic glycosides usually have a bitter taste [11]; thereby a distinction between sweet and bitter cassava varieties is made. Hydrogen cyanide content of sweet cassava (low cyanide varieties) will typically range from 15 to 50 mg/kg on a fresh weight basis, and can be processed adequately by peeling and cooking (e.g. roasting, baking or boiling) [14]. Bitter varieties of cassava (high cyanide content) require more extensive processing to remove the cyanogenic potential; therefore those varieties are not normally commercially traded [14].

Glycosides are present in all parts of cassava plants, but leaves contain approximately 10% more linamarin than roots [14]. Young leaves may contain around 600 mg HCN/kg of fresh material [13], thus requiring blanching and extensive boiling before consumption.

The safe level of cyanide in cassava flour has been set to 10 ppm [15]. This level is much lower than that present in many varieties of cassava. Thus, the widespread use of cassava roots as food requires careful treatment and training in cooking, since it may increase public health risks.

Hydrogen cyanide toxicity is derived from its ability to bind to the iron contained in the enzyme cytochrome oxidase, inhibiting its activity and disrupting the electron transport chain in the mitochondria. The inhibition of this metabolic reaction impedes the synthesis of adenosine triphosphate (ATP, the way cells store energy) leading to a rapid death of the cell tissue.

Acute cyanide poisoning includes diverse symptoms such as dizziness, headaches, general weakness, stomach pains, vomiting, diarrhea and mental confusion, among others. When the concentration of hydrogen cyanide exceeds the limit that an individual is able to detoxify or tolerate, cyanide poisoning is mortal. It was reported that the acute lethal dose of hydrogen cyanide for humans is 0.5 to 3.5 mg / kg body weight [14].

A chronic ingestion of sub-lethal concentrations of cyanide can also trigger long-term consequences in human health. Particularly, linamarin, the main cyanogenic glycoside present in cassava, has been associated to neuronal damage in structures of the central nervous system involved in motor processing [16]. One of the long-term diseases related to dietary cyanide intake is "konzo", an upper motor neuron disease characterised by the sudden onset of irreversible but non-progressive symmetric spastic paraparesis (difficulty in the movement of the lower limbs).

Other pathology related to the regular consumption of dietary cyanide is the tropical ataxic neuropathy (TAN), a term used to describe several syndromes attributed to toxico-nutritional neuropathies, whose most common symptoms are of sensory type, initially confined to the legs, with an ataxic gait (impaired coordination of voluntary movements and balance) and less constant visual impairment [14; 17].

In humans, the detoxification process triggered by the ingestion of cyanide is produced by the enzyme rhodanese. In this process, which takes place predominantly in the liver, rhodanese use sulphur donors for converting cyanide to thiocyanate, which is transported through the bloodstream to the kidneys, where it is eliminated in the urine.

Therefore, the nutritional status of the individual, particularly the adequate intake of sulphur containing amino acids methionine and cysteine, is crucial in the overall toxicity produced by cyanide ingestion.

Thiocyanate produced with this detoxification pathway is not completely innocuous, because it inhibits the capture of iodine in the thyroid, causes increased secretion of thyroid stimulating hormone and also reduces the levels of thyroxine hormone, involved in growth and development. Therefore, cyanide intake in the diet also brings problems of goiter and cretinism.

Besides cyanogenic glycosides, other less hazardous antinutrients (factors that interfere with the absorption of certain nutrients) such as tannins and phytic acid are also present in cassava, although none of them have been reported to be in concentrations that are of concern.

Concerning other antinutritional factors affecting nutritional quality of plant products, phytate, the anion released from phytic acid dissociation, is typically related to the reduction of the uptake of iron and is also implicated in the precipitation of other elements such as magnesium, zinc, copper, calcium and manganese. On the other hand, it has been reported that phytate ion exhibits anticarcinogenic activity in cell culture and animal model assays [12].

Phytic acid content of cassava flour has been informed to be in the range of 95 to 136 mg/ 100g on dry basis [18], approximately 10 times lower than that reported for sweet potato flour [19].

Tannins, for its part, act as anti-thiamine factors by destroying thiamine (vitamin B1), thus reduce its absorption in the organism. Tannins also bring astringency to foods due to their interaction with salivary proteins. On the other side, condensed tannins have also been reported to reduce the risk of coronary diseases [20-21]. Oboh et al., [22] informed tannins content in cassava flour of 200 mg/100 g, similar to that found in other edible roots and tubers flours [12].

Cassava Flour

Cassava tuberous roots can be submitted to relatively simple procedures such as grating or slicing and subsequent drying to obtain flour, being this derivative a naturally gluten-free product. This low-tech process can be carried out by farming communities and cooperatives because it requires no equipment or invest in high-cost facilities, thus contributing to the development of economically depressed regions. The yield of this procedure is around 270 g flour per kg fresh root. In an own work [23], it was found that the obtaining procedure did not affect flour yield.

Chemical composition analysis indicated that cassava flour contains 92.5±0.9% (dry basis) of carbohydrates, being 83.5±2.5% starch. Besides, 10.2±0.1% corresponds to moisture, 3.45±0.04% to ashes, 0.48±0.02% to lipids, 3.7±0.7% to proteins and 1.9±0.2% to acid detergent fiber. According to Shittu et al., [24], starch, sugar, amylase, and the pH of cassava flour differed significantly among cassava clones. The ash, sugar, and protein contents varied slightly between 1.5–3.6%, 1.2–2.2%, and 0.8–1.9% (dry basis), respectively. The flours cyanogenic potential (CNP) also ranged between 0 and 38 mg kg-1 (dry basis). The starch and amylose contents showed the highest variation, ranging from 65–88% and 13–23% (dry basis), respectively [24-26].

With regard to mineral content, Oboh et al., [27] stressed that cassava flours contains 43.8±0.3% (dry basis) of sodium, 49.8±0.4% of potassium, 61.6±0.7% of calcium, 43.4±0.2 of magnesium and 26.0±0.4% of iron.

Morphologically, cassava flours were characterized by the presence of starch granules as well as irregular, deformed, truncated particles with low sphericity [23], which could be mainly associated to protein aggregates and fiber components. Figure 3 shows scanning electron microscopy (SEM) micrographs of cassava flours obtained by slicing (Figure 3a) as well as its starch (Figure 3b). The flour obtaining procedure did not affect SEM observations.

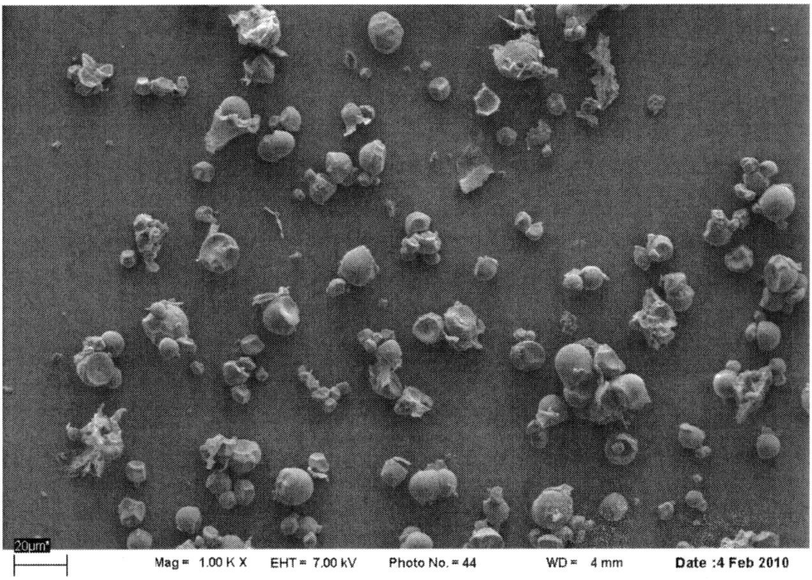

a

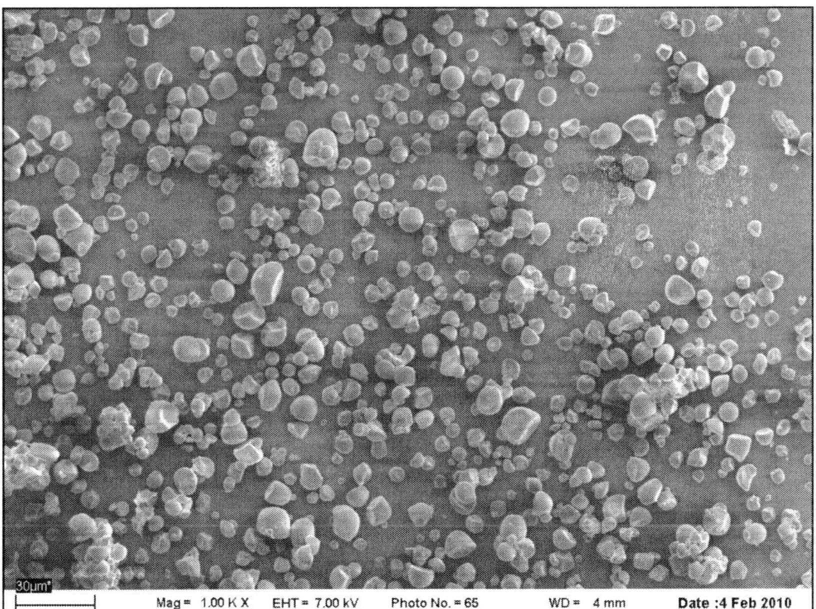

b

Figure 3. SEM micrographs of a- cassava (*Manihot esculenta*) flour and b- cassava starch. Magnification used is indicated in the micrograph.

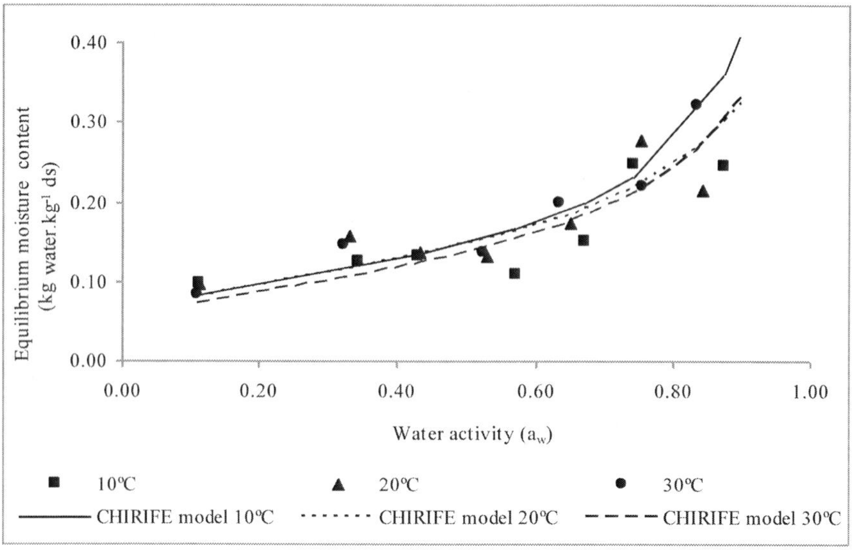

a

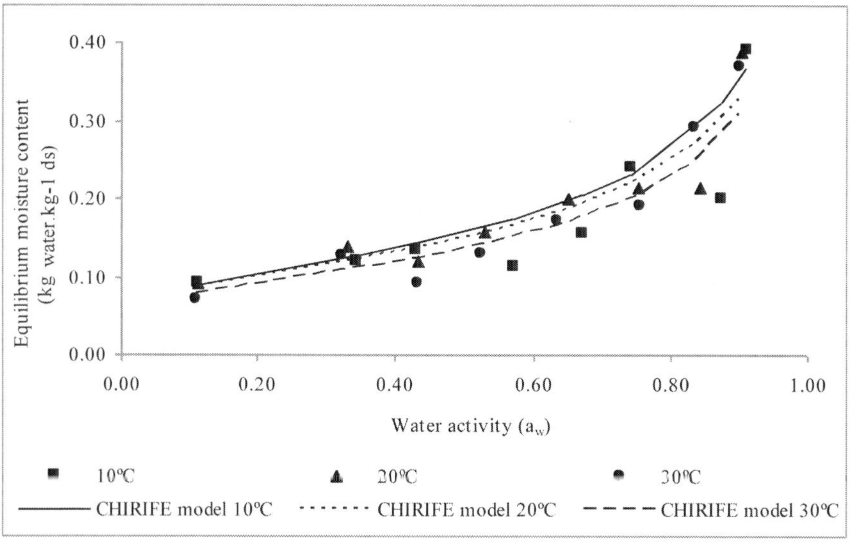

b

Figure 4. (Continued)

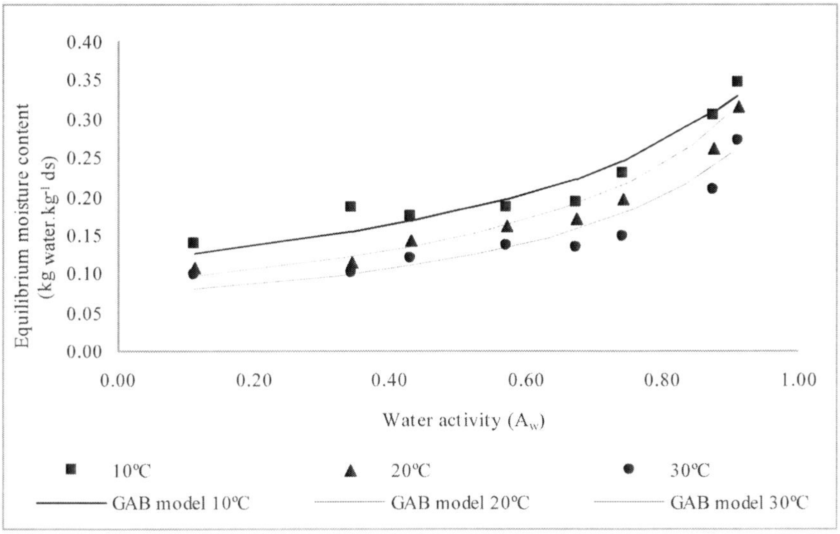

c

Figure 4. Experimental data of equilibrium moisture content (dry basis) as a function of aw at 10, 20 and 30°C and their corresponding fitting to the Chirife model for: a- cassava flour obtained by slicing (CFS); b- cassava flour obtained by grating (CFG) and c- cassava starch.

Differential scanning calorimetry (DSC) thermograms corresponding to cassava flours showed a single thermal event associated with the gelatinization of starch. Gelatinization temperature of flour was higher than that of its starch. This trend was also reported by Jane et al. [28] for flours and starches of different botanical origin. The values of onset temperature (70.1±0.8°C), gelatinization temperature (75.7±0.9°C) and enthalpy (10.5±0.9 J/g) found for cassava flour were consistent with those reported by Charoenkul et al., [25], particularly for cassava varieties that were classified by these authors as belonging to ''mealy and firm group'', according to the type of texture of cooked roots.

Likewise, sorption isotherms of cassava flour were determined at 10, 20 and 30°C. Experimental data were satisfactorily fitted using different mathematical models, obtaining the best fitting with that proposed by Chirife (Figure 4). Also, from GAB (Guggenheim, Anderson and de Boer) model thermodynamic parameters associated with water adsorption process can be calculated, being the monolayer water content (Xm) indicative of product stability. Labuza [29] have stressed that the maximum monolayer moisture content for foods corresponds to 10% (dry basis), because when this value is exceeded, the stability of the product is compromised. All samples could be considered as products with an acceptable stability, being the Xm values of cassava flours 0.082 g water/g dry solid and 0.088 g water/g dry solid that corresponding to its starch at 20°C. Besides, Xm tended to decrease with increasing temperature. This phenomenon could be attributed to the reduced availability of active sites for water adsorption [25; 30].

Cassava flour is mainly used in baking and confectionery products to substitute wheat flour at different levels as well as the application in the manufacture of weaning foods and pasta. Likewise, starch production can be destined to the food, pharmaceutical, and chemical industries [24].

Many recipes in South America and the Caribbean are made of cassava flour. For example, "Casabe" is the unleavened bread made of cassava flour which was originally produced by American natives, principally from the Caribbean. As wheat bread, it can be consumed alone or accompanying other meal. Jamaicans have a variant of this recipe, known as "Bammy", which is thicker bread than Casabe and the preparation is usually soaked in coconut milk before cooking.

Other bakery product made of cassava flour is the "cassava cheese bread", very popular in some countries of South America. Small variations on the recipe created characteristic products with different names according to the region of origin. In Brazil, those cheese buns are known as "pão do queijo", in Bolivia are called "cuñapé" and in Paraguay and Argentina they are commonly known as "chipa" (Figure 5).

Other food uses of cassava include the fermentation for production of cassava beer ("Impala®", the first commercial-scale cassava-based beer, was launched in 2011 in Mozambique) and "cassareep", a syrup obtained from the caramelization of cassava juice after boiling (originated by the Guyanese Amerindians). Furthermore, other applications such as medicinal use of modified cassava starch as a promoter for fluid and electrolyte absorption after diarrhoea episodes have also been tested [31].

The diffusion of the different uses of cassava as food ingredient, the potential of its derived products as raw materials or additives for manufactured foods as well as its nutritional quality could function as promoters for spreading its production. Some of the factors limiting its use as ingredient for homemade bakery products are the distinct rheological behavior and cooking properties with respect to those of wheat flour, which makes necessary the training of new users for its proper implementation in cooking.

Figure 5. Cassava cheese bread "chipa".

CASSAVA STARCH

Cassava starch (commonly known as "tapioca") is the main product obtained from the roots and can be easily extracted by a rather simple technical process (Figure 6). Starch yield depends on cassava cultivar, as it is shown in Table 2 for those cultivated in the north-east region of Argentina, east of Paraguay and south of Brazil. Starch yield is a relevant selection criterion in breeding programs. Besides, amylose content was also affected by the cultivar analyzed (Table 2).

SEM analyses of cassava starch granules morphology showed a bimodal distribution with two distinct populations of granules centred at about 3 and 10 μm (Figure 3b). Sriroth et al., [32] have stressed that the size distribution of cassava starch granules depends on the crop phase at harvest time. Likewise, granule shape, size and surface are simple physical characteristics which have an impact on starch functionality.

Technological Properties

Cassava starch showed a water retention capacity of 88.2±1.6%, lower than that of other ingredients or food products, such as dietary fibers from oat bran (2.10 g of water / g of additive material dry), rice bran (4.89 g water / g dry matter), soy flour (4.79 to 6.75 g water / g dry matter) and wheat bran (5.03 g of water / g dry matter), [33-35].

Figure 6. Cassava starch produced in Misiones (Argentina).

Table 2. Extraction yield, amylose content and thermal properties of cassava starch obtained from different cultivars

Cassava cultivar*	Starch extraction yield (%)	Amylose content (%)	Thermal properties	
			Gelatinization temperature (°C)	Enthalpy (J/g, dry basis)
Pomberí	82.39	13.50±0,31	66.88 ± 0.08	13.92 ± 0.61
IAPAR 989	86.12	12.37±0,47	64.67 ± 0.25	14.71 ± 1.53
Verde Olivo	94.33	14.20±0,51	67.05 ± 0.34	14.69 ± 0.70
Bahianinha	84.63	12.68±0,98	67.59 ± 0.76	13.90 ± 1.90
CA 25-1	78.43	12.80 ± 0.77	67.67 ±0.71	13.74 ± 1.06
CA 145	81.81	13.14 ± 0.73	67.73 ± 0.75	11.28 ± 1.34
IAC 90	70.81	18.06 ± 0.90	69.16 ± 0.41	13.37 ± 0.40
Rito B	79.67	11.67 ± 0.27	69.27 ± 0.49	11.90 ± 0.76

* Cassava cultivars grown in the north-east region of Argentina, east of Paraguay and south of Brazil.

On the other hand, the swelling capacity of the granules plays an important role in the rheological behavior of starch suspensions and can be determined according to a modification of the method of Tsai et al., [36]. Granules swelling capacity increases with rising in heating temperature; this is attributed to the replacement of interactions between chains with H-bonds interactions with the absorbed water molecules [37].

Aqueous suspensions of the studied starches exhibited DSC thermograms with a single endothermic transition, typical of the gelatinization process of high water content suspensions. Thermal properties of starches depended on botanical source, being cassava gelatinization temperature significantly ($p<0.05$) lower than that of cereal and legume such as corn and wheat, indicating a better aptitude for cooking [38-39]. Cassava cultivar slightly affects gelatinization temperature (Table 2). Gelatinization enthalpy, which is an indicator of the molecular order within the granule, was similar to those informed by Betancur et al., [40].

Cassava starch pastes presented a pseudoplastic behavior, which was satisfactorily adjusted by Ostwald de Waele model. Besides, gelatinized cassava starch suspensions exhibited a high thixotropy (7672 Pa/s) and apparent viscosity values (166.50±5.70 mPa s) indicating that the remaining amylopectin inside starch granules reinforces amylose network. These rheological properties are important in food formulations since a pseudoplastic material has suspending properties at low shear rates and its viscosity becomes sufficiently low when it is processed at higher shear rates [41]. Likewise, thixotropic systems exhibit decreasing shear stress and apparent viscosity over time at a fixed shear rate.

Mechanical spectra obtained by oscillatory assays in a rheometer (RheoStress 600, Haake, Germany) showed that cassava starch gelatinized suspensions presented a typically gel behavior since significantly higher elastic module values than the viscous ones (G′>G”) were registered over almost all the frequency range assayed, and curves maintained practically parallel. Lee and Yoo [42] also reported a weak gel-like behavior for sweet potato starch pastes.

Retrogradation is a process that occurs when the amylose and amylopectin chains from the starch pastes are re-associated to form a more ordered structure. When pastes were stored for 11 days under refrigeration (4°C), the typical endothermic peak located at 45-55°C was not observed by DSC, indicating the low tendency of cassava starch pastes to retrogradate. Moreover, a low syneresis degree was observed (4.07±0.77%, respectively) [43]. Changes in the texture of starch pastes during storage should be evaluated since they are closely related to its applications in the formulation of foods [44]. In the mechanical profile obtained in a Texture Analyzer (TA xT2i, United Kingdom) fracturability peak was not observed. This peak is attributable to pastes stiffness increased due to the retrogradation of the amylopectin. Thus texture profile analysis (TPA) results correlate with those obtained from DSC assays. The hardness, gumminess and chewiness of pastes decreased significantly ($p<0.05$) during storage, while the remaining TPA parameters were not affected (Table 3).

Table 3. Texture profile analysis (TPA) parameters of cassava starch pastes stored at 4°C

TPA parameter	1 day of storage	11 days of storage
Fracturability (N)	nd	nd
Hardness (N)	0.34±0.01	0.28±0.02
Elasticity	1.00±0.00	1.00±0.02
Cohesivity	0.70±0.01	0.66±0.03
Gumminess (N)	0.24±0.01	0.19 ± 0.02
Adhesiveness (N mm)	0.52±0.18	0.19 ± 0.02
Chewiness (N)	0.24±0.01	0.19 ± 0.02

nd: not detected.

Furthermore, it is important to stress that the comparison of the texture parameters of starch pastes with other reported data is difficult since they depend on several factors: the concentration of starch suspension, storage temperature and time as well as other experimental conditions.

Nutritional Aspects

Cassava starch was analyzed in terms of rapidly digestible, slowly digestible, and resistant starch fractions (RDS, SDS, and RS, respectively), obtained from gelatinized starch hydrolysis.

Digestion kinetics was performed with pancreatic α-amylase (20 Uml-1) and amyloglucosidase (15 Uml-1) and the concentration of released glucose after 20 and 120 min incubation was analyzed. The starch fraction digested at 20 min was classified as RDS, the one hydrolyzed between 20 and 120 min was the SDS and the remaining non-digested starch after 120 min was designate as RS [45-46].

This RS is classified as RS3, which appertain to the non-hydrolyzed fraction mainly composed of retrograded amylose formed during the cooling process of gelatinized starch pastes [47].

The hydrolysis index (HI) of native and gelatinized cassava starches was calculated following the protocol described by Goñi et al., [48], by comparison of the percentage of total starch hydrolyzed obtained from the hydrolysis curve (0–180 min) with alpha amylase and that released from white bread over the same hydrolysis period (Figure 7). Hydrolysis curve was satisfactorily fitted to the model proposed by Goñi et al., [47-48], and the obtained HI values were used to estimate the glycemic index (GI).

GI is an indicative of the amount of sugar released into the bloodstream after ingestion of starchy foods, therefore associated to the amount of energy the food can provide.

A relationship between the GI in vivo and the amount of starch hydrolyzed in vitro after incubation for 180 min has been established [48], being a GI of 94 designated to white bread [49].

The studied cassava gelatinized starch gave total hydrolysis percentages above those corresponding to white bread in all sampling points analyzed, leading to HI and GI values of 136.69±0.16 and 114.75±0.09 respectively [47]. Ingredients with estimated GI values higher than that from white bread could be useful for the formulation of dietary products with high energy contents.

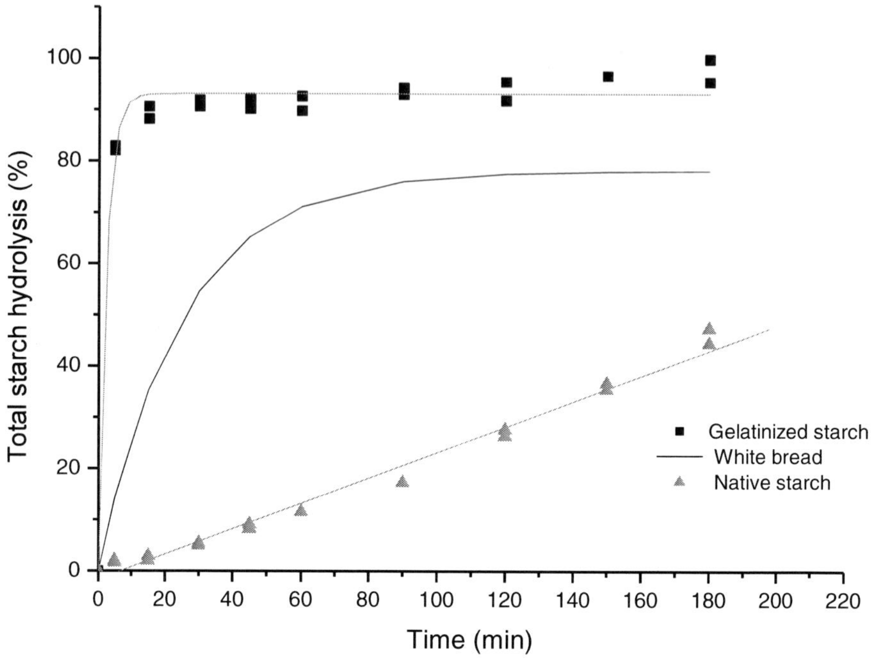

Figure 7. Enzymatic hydrolysis kinetics (total starch hydrolysis percentage) of native and gelatinized cassava starch and the respective curve from white bread; symbols indicate experimental data and lines represent the fitting curves to the model proposed by Goñi et al. [48].

This feature is very promising for the development of cassava starch based food, not only for people with celiac disease but also for infants and groups with special requirements such as high performance athletes. It is worth to mention that cassava is a rich source of carbohydrates whose content ranges from 80-86% [25].

As expected, the GI of gelatinized starch was higher than that of native starch (56.2±0.58) as observed in Figure 7. This is due to the disruption of the hydrogen bonds between starch chains during the gelatinization process which facilitates the enzymatic attack [50]. Due to the slower digestion kinetics of native starch, maximum percentage of starch hydrolysis was not achieved after 3h of digestion (Figure 7); therefore long-term (20.5 h) hydrolysis kinetics of raw native starch was assayed.

Long-term incubation with pancreatic α-amylase showed a maximum percentage of enzymatic digestion of 86.0% (dry basis) after 8 h, similar to that reported by Srichuwong et al., [51] for type A cassava starch, implying that the remaining 14.0 % corresponded to type II resistant starch (RS2, resistant starch that occurs in its natural granular form), based on the classification derived from the nutritional characteristics of starch [46].

Some beneficial health effects have been described for RS [52]. RS may be classified as a component of dietary fiber, due to its high resistance to mammalian enzyme attack. RS physiological activities related to health improvement resemble to those of soluble fiber, presenting, as most relevant features, a positive impact on colonic health and tumorigenesis, and reducing serum cholesterol and triglycerides [52].

ADDING VALUE TO BY-PRODUCTS

Cassava bagasse is a by-product of the cassava-processing industry, which is a fibrous material that contains about 30-50% starch on dry weight basis [53]. It is considered an attractive raw material, not only as a source of fiber but also for the manufacture of products derived from bacterial or fungal solid state fermentation due to its relatively high starch content. Furthermore, the use of cassava bagasse for microbial mediated bioconversion process is advantageous with respect to other crop wastes because of its low ash content [53].

Implementation of cassava bagasse fermentation has been proposed for obtaining non-food products, such as pullulan [54] or n-butanol [55] and food-related products such as nutritionally improved animal feed [56] or aroma compounds [53; 57-58].

The use of fiber obtained from cassava bagasse has been studied as natural filler intended to reinforce biomaterials and modify their hydrophobicity [59-61].

Fiber derived from cassava bagasse is also attractive as food additive. Dietary fiber (DF) is the collective name given to all components of plant cell walls that are not hydrolyzed by enzymes in the digestive tract. The DF contains numerous reactive groups able to interact with bile acids, preventing their absorption into the intestinal lumen and thus reducing the plasma cholesterol level [62]. Dietary fiber has also known beneficial effects on regulation of intestinal function, the prevention of colon disorders and the regulation of some metabolic diseases. Furthermore, DF also helps to control diabetes and has been proved to prevent the onset of diverticula [63]. Therefore, the use of dietary fiber enriched food products is taking increasing attention for the food industry, particularly by exploiting fiber-rich by-products of the agricultural processing industry [64].

Cassava as a Gluten-free Ingredient

One of the main potential benefits of cassava as ingredient for bakery products is the fact that is naturally gluten free, which poses it as a food product plausible for being incorporated into celiac disease people's diet.

Celiac disease (CD) is an autoimmune disorder characterized by damage in the small intestine tissue due to an atrophy of villi lining. This tissue damage is originated by an immune reaction to gliadins (the predominant protein present in gluten) [65]. Nutritional deficiencies are often present in patients with CD due to a reduced capability of the small intestine to absorb vitamins and other nutrients. Furthermore, untreated celiac disease also leads to the development of anaemia [65].

According to the World Gastroenterology Organisation Global Guidelines [66], the worldwide prevalence of celiac disease has been reported to be of 1% of world population, significantly higher than that recognized 20 years ago. It was also reported that the number of cases of CD is rising in North America and Europe. This increased prevalence is probably related to the recognition of wide clinical variability in the disorder. Moreover, the real number of cases in the world is expected to be much higher than that found up to now since it has been informed that there are far more undiagnosed cases than diagnosed cases. It is believed that the ratio of diagnosed to undiagnosed cases of CD varies greatly between countries (1: 2 in Finland, 1: 20 in Argentina and United States) [66].

It is well known that so far the only treatment for celiac disease is a lifelong adherence to a gluten-free diet. The long term dietary habits and foods choices of celiac patients subjected to a strict gluten-free diet usually result in an unbalanced intake of carbohydrates, protein, and fat, as well as limited intake of certain essential nutrients [12]. It is well recognized that white bread, among other starchy foods, belongs to the base of the food pyramid. It is one of the most widely spread foods present in non-restricted eating habits, providing complex carbohydrates which represent an important source of energy, especially in low-fat diets. Therefore, the replacement of white bread in the diet of celiac patients is essential for maintaining an adequate intake of carbohydrates.

Scientific community is making efforts for mimicking the cohesiveness of wheat flour dough on gluten free bakery products. The mixture of gluten-free flours (principally rice, corn, soy, and buckwheat flours) and starches from different botanical sources is one of the strategies proposed for compensating the absence of gluten. In that sense, cassava starch is being widely studied as enhancer of dough quality of gluten-free flours from other origins. In the work of Onyango et al., [67], pregelatinized and native cassava starches were studied as additives for improving rheological and textural properties on sorghum-based butters and breads. It was observed that the addition of native starch increased batters fluidity, whereas raising the amount of pregelatinized cassava starch increased batters firmness, giving dough-like consistencies.

In order to introduce further improvements on making gluten-free batters and to simulate the distinctive characteristics of wheat dough, the use of certain additives to gluten-free flours and starches mixtures is also being explored. Hydrocolloids, for example, have been assayed to improve dough development, water binding and gas retention. The addition of xanthan gum, carboxymethylcellulose, carrageenan, alginate and high methoxy hydroxypropyl methylcellulose (HM HPMC) have been tested on formulations based on rice flour and cassava starch, but the solely addition of these hydrocolloids seems not to be enough to effectively simulate gluten dough and wheat bread properties [68-69]. Similar results were obtained using emulsifiers as optimizers for rice and cassava gluten free dough [69].

On the other hand, the incorporation of proteins isolates from different sources to gluten-free formulations has also been studied as potential strategy to simulate the protein network formed by gluten on wheat dough, which gives its characteristic cohesiveness and elasticity and provides a good mouthfeel of bread crumb. Crockett et al., [70] studied the effect of the addition of soy protein isolate and egg white solids on the physicochemical properties of HPMC added gluten free dough, primarily made of rice flour and cassava starch. Only egg white solids at highest concentration assayed (15%) gave an improvement on loaf volume, the other conditions assayed showed a detrimental activity of the interaction of added proteins and HPMC to loaf quality [70]. As expected, the more complex the formulation is, the more difficult becomes to predict the results to be obtained on batter and bread qualities. On a further attempt to try mimicking the protein network formed in gluten dough, the addition of protein polymerizing enzymes on gluten-free mixtures has also been investigated. In the research work of Onyango et al., [71], the addition of transglutaminase on pregelatinized cassava starch, sorghum and egg white batter resulted in and increment of batter strengthening due to cross-linking action. The effect of transglutaminase and pregelatinized cassava starch to jasmine rice flour was as well effective on improving jasmine rice bread quality [72]. Glucose oxidase was also tested on rice flour and cassava starch formulation but no improvement was observed in that case [69]. It is worth to mention that enzyme activity highly depends on the substrate, thus obtained results are expected to vary significantly on different flour/starch mixtures.

Besides its potentiality as GF ingredient for bakery products, cassava is also a promising raw material for other types of celiac grade foods such as pasta. In the work of Fiorda et al., [73], quality attributes of gluten-free pasta made of amaranth flour, cassava starch and cassava bagasse were assessed and compared to those of commercial semolina pasta. Authors found a formulation with quality attributes of color, texture and nutritional value comparable to those of commercial wheat products. The use of cassava bagasse includes an additional benefit on this new product since it represents a healthier option, due to increased fiber content, which may also be chosen by non celiac consumers. Furthermore, the exploitation of this sub-product may help to increase the profitability of cassava starch industries [73] and simultaneously reduce the problem of solid waste disposal.

References

[1] FAOSTAT 2014 Online database. Available in http://faostat3.fao.org/faostat-gateway/go/to/download/Q/*/E. Date accessed February 15, 2014.

[2] RWF 2014 Cassava factory (online). Date accessed February 26, 2014. Available from: http://www.ruralwf.org/node/9.

[3] K. Kawano, *Journal of Root Crops* 37, 101 (2011).

[4] FAO 2006 Starch market adds value to cassava (in spanish) (online). Date accessed February 28, 2014. Available from: http://www.fao.org/ag/esp/revista/0610sp1.htm.

[5] R. M. Welch, and R. D. Graham, *J. Exp. Bot.* 55, 353 (2004).

[6] C. Iglesias, J. Mayer, L. Chavez, et al., *Euphytica* 94, 367 (1997).

[7] A. L. Chavez, J. M. Bedoya, T. Sánchez, et al., *Food & Nutrition Bulletin* 21, 410 (2000).

[8] H.-H. Yeoh, and V.-D. Truong, *Journal of the Science of Food and Agriculture* 70, 51 (1996).

[9] H. Ceballos, T. Sánchez, A. L. Chávez, et al., *Journal of Food Composition and Analysis* 19, 589 (2006).
[10] J. H. Bradbury, and I. C. Denton, *Food Chemistry* 123, 840 (2010).
[11] H. P. Makkar, P. Siddhuraju, and K. Becker, *Plant secondary metabolites*, (2007).
[12] C. Dini, M. A. Garcia, and S. Z. Viña, *Food & Function* 3, 606 (2012).
[13] F. F. Teles, *Food Nutr. Bull.* 23, 407 (2002).
[14] FSANZ, *Cyanogenic glycosides in cassava and bamboo shoots*, Food Standards Australia New Zealand, Canberra BC (2004).
[15] FAO/WHO, *Joint FAO/WHO food standards programme. In* Codex Alimentarius Commission XII. Suppl. 4. Rome (1991).
[16] E. Rivadeneyra-Dominguez, A. Vazquez-Luna, J. F. Rodriguez-Landa, et al., *Food Chem. Toxicol.* 59, 230 (2013).
[17] J. Cliff, H. Muquingue, D. Nhassico, et al., *Food Chem. Toxicol.* 49, 631 (2011).
[18] A. L. Charles, K. Sriroth, and T.-c. Huang, *Food Chemistry* 92, 615 (2005).
[19] E. N. Eluagu, and I. A. Onimawo, *Electronic Journal of Environmental, Agricultural and Food Chemistry* 9, 1000 (2010).
[20] M. Dell'Agli, A. Busciala, and E. Bosisio, *Cardiovasc. Res.* 63, 593 (2004).
[21] P. Gresele, C. Cerletti, G. Guglielmini, et al., *J. Nutr. Biochem.* 22, 201 (2011).
[22] G. Oboh, and A. A. Akindahunsi, *Food Chemistry* 82, 599 (2003).
[23] M. C. Doporto, C. Dini, A. Mugridge, et al., *Journal of Food Engineering* 113, 569 (2012).
[24] T. A. Shittu, L. O. Sanni, S. O. Awonorin, et al., *Food Chemistry* 101, 1606 (2007).
[25] N. Charoenkul, D. Uttapap, W. Pathipanawat, et al., *LWT - Food Science and Technology* 44, 1774 (2011).
[26] F. N. A. Aryee, I. Oduro, W. O. Ellis, et al., *Food Control* 17, 916 (2006).
[27] G. Oboh, A. A. Akindahunsi, and A. A. Oshodi, *Journal of Food Composition and Analysis* 15, 617 (2002).
[28] J. Jane, L. Shen, J. Chen, et al., *Cereal Chemistry* 69, 528 (1992).
[29] T. P. Labuza, *Moisture Sorption: Practical Aspects of Isotherm Measurement and Use*, American Association of Cereal Chemists, St. Paul, MN. (1984).

[30] M. C. Doporto, A. Mugridge, M. A. García, et al., *Food Chemistry* 126, 1670 (2011).
[31] M. A. Wingertzahn, S. Teichberg, and R. A. Wapnir, *Pediatr. Res.* 45, 397 (1999).
[32] K. Sriroth, V. Santisopasri, C. Petchalanuwat, et al., *Carbohydrate Polymers* 38, 161 (1999).
[33] A. Abdul-Hamid, and Y. S. Luan, *Food Chemistry* 68, 15 (2000).
[34] C. Chen, *A study of equilibrium relative humidity for yellow dent corn kernels.*, University of Minnesota, St. Paul, MN. (1988).
[35] A. Heywood, D. Myers, T. Bailey, et al., *Journal of the American Oil Chemists' Society* 79, 1249 (2002).
[36] M.-L. Tsai, C.-F. Li, and C.-Y. Lii, *Cereal Chemistry Journal* 74, 750 (1997).
[37] A. C. Eliasson, *Starch in Food: Structure, Function and Applications*, Taylor & Francis, (2004).
[38] J.-Y. Li, and A.-I. Yeh, *Journal of Food Engineering* 50, 141 (2001).
[39] E. A. Melo, T. L. Stamford, M. P. Silva, et al., *Bioresour. Technol.* 89, 103 (2003).
[40] D. A. Betancur, L. A. C. Ancona, R. I. Guerrero, et al., *Starch - Stärke* 53, 219 (2001).
[41] L.-m. Che, D. Li, L.-j. Wang, et al., *Carbohydrate Polymers* 74, 385 (2008).
[42] H.-L. Lee, and B. Yoo, *Starch - Stärke* 61, 407 (2009).
[43] O. V. López, *Desarrollo, caracterización y aplicación de envases biodegradables a partir de almidón.*, National University of La Plata, La Plata (2011).
[44] K. S. Sandhu, and N. Singh, *Food Chemistry* 101, 1499 (2007).
[45] J. Juansang, C. Puttanlek, V. Rungsardthong, et al., *Food Chemistry* 131, 500 (2012).
[46] H. N. Englyst, S. M. Kingman, and J. H. Cummings, *Eur. J. Clin. Nutr.* 46 Suppl 2, S33 (1992).
[47] M. C. Doporto, C. Dini, S. Z. Viña, et al. Preprint, *Starch - Stärke* (2013). DOI 10.1002/star.201300196.
[48] I. Goñi, A. Garcia-Alonso, and F. Saura-Calixto, *Nutrition Research* 17, 427 (1997).
[49] B. C. Ebah-Djedji, K. M. Dje, B. N'Zue, et al., *Pakistan Journal of Nutrition* 11, 414 (2012).
[50] H.-J. Chung, H. S. Lim, and S.-T. Lim, *Journal of Cereal Science* 43, 353 (2006).

[51] S. Srichuwong, T. C. Sunarti, T. Mishima, et al., *Carbohydrate Polymers* 60, 529 (2005).
[52] M. G. Sajilata, R. S. Singhal, and P. R. Kulkarni, *Comprehensive Reviews in Food Science and Food Safety* 5, 1 (2006).
[53] A. Pandey, C. R. Soccol, P. Nigam, et al., *Bioresource Technology* 74, 81 (2000).
[54] K. R. Sugumaran, P. Jothi, and V. Ponnusami, *Carbohydrate Polymers* 99, 22 (2014).
[55] C. Lu, J. Zhao, S. T. Yang, et al., *Bioresour. Technol.* 104, 370 (2012).
[56] C. Sriherwanto, C. Koob, and B. Bisping, *New Biotechnology* 25, Supplement, S289 (2009).
[57] P. Christen, A. Bramorski, S. Revah, et al., *Bioresource Technology* 71, 211 (2000).
[58] A. Bramorski, P. Christen, M. Ramirez, et al., *Biotechnology Letters* 20, 359 (1998).
[59] E. d. M. Teixeira, D. Pasquini, A.A.S. Curvelo, et al., *Carbohydrate Polymers* 78, 422 (2009).
[60] D. Pasquini, E. d. M. Teixeira, A. A. d. S. Curvelo, et al., *Industrial Crops and Products* 32, 486 (2010).
[61] E. d. M. Teixeira, A. A. S. Curvelo, A. C. Corrêa, et al., *Industrial Crops and Products* 37, 61 (2012).
[62] Z.-Y. Chen, K. Y. Ma, Y. Liang, et al., *Journal of Functional Foods* 3, 61 (2011).
[63] K. A. Steinmetz, and J. D. Potter, *Journal of the American Dietetic Association* 96, 1027 (1996).
[64] M. Elleuch, D. Bedigian, O. Roiseux, et al., *Food Chemistry* 124, 411 (2011).
[65] T. Griga, *Celiac disease*. *In* Autoimmune Diagnostics. De Gruyter, Berlin (2012).
[66] J. C. Bai, M. Fried, G. R. Corazza, et al., *Celiac disease*, World Gastroenterology Organisation Global Guidelines, (2012).
[67] C. Onyango, C. Mutungi, G. Unbehend, et al., *LWT - Food Science and Technology* 44, 687 (2011).
[68] R. Crockett, P. Ie, and Y. Vodovotz, *J. Food Sci.* 76, E274 (2011).
[69] L. S. Sciarini, P. D. Ribotta, A. E. León, et al., *Journal of Food Engineering* 111, 590 (2012).
[70] R. Crockett, P. Ie, and Y. Vodovotz, *Food Chemistry* 129, 84 (2011).
[71] C. Onyango, C. Mutungi, G. Unbehend, et al., *Journal of Food Engineering* 97, 465 (2010).

[72] W. Pongjaruvat, P. Methacanon, N. Seetapan, et al., *Food Hydrocolloids* 36, 143 (2014).
[73] F. A. Fiorda, M. S. Soares Jr, F. A. da Silva, et al., *LWT - Food Science and Technology* 54, 132 (2013).

In: Cassava
Editor: Francis P. Molinari
ISBN: 978-1-63321-031-8

Chapter 5

CASSAVA BREAD (CASABE) IN VENEZUELA: SOME NUTRITIONAL CONSIDERATIONS

Benito Infante [*] ***and Omar Garcia***
Laboratorio de Investigaciones, Escuela de Nutrición y Dietética, Facultad de Medicina, Universidad Central de Venezuela
Docente-Investigador, PhD, Departamento de Ciencias Básicas, Cátedra de Bioquímica, Current Director of the Escuela de Nutrición y Dietética, Facultad de Medicina, Universidad Central de Venezuela
Docente-Investigador, MSc, Cátedra de Ciencias y Tecnología de Alimentos, Departamento de Nutrición y Alimentación, Escuela de Nutrición y Dietética, Facultad de Medicina, Universidad Central de Venezuela

ABSTRACT

Cassava (*Manihot esculenta Crantz*) is a tuber of the Family Euphorbiaceae of the genus Manihot originary from South America. Is a popular edible root since centuries in many parts of Africa, Asia and South America. Cassava is an important energy sources for tropical areas of the world, rich in calories but deficient: in protein, fats, minerals and vitamins and contains in low proportions, poliphenols, tannins and cyanogenic compounds.

[*] E-mail: infanterster@gmail.com.

Geographical localization of cassava bread supplier: Estado (Edo) Sucre, Venezuela east, “casaba Sucre”, Edo Bolivar South of venezuela “casaba Amazon” and Edo Miranda, Venezuela Center “casaba Miranda”. Using the enzymatic method of Prosky, was determined total dietary fiber(TDF), soluble dietary fiber(SDF) and insoluble dietary fiber (IDF) for “casaba Sucre” 5.4 %, 0.8 % and 4.6 % respectively; “casabe Miranda” 5.2 %, 0.7 and 4.5 % respectively; “casabe Amazon” 6.3 %, 1.6 % and 4.7 % respectively. Previous studies have reported average values for these varieties, IDF 4.92 - 5.67 % and SDF 3.4 - 3.78 % using the Asp method. The test for starch digestibility of “casabe” by pancreatic amylase showed to be very similar to the wheat starch thermally treated, however its rate of *in vitro* digestibility was relatively low 73 %. Studies in the rat model (Sprague Dawley) using adult males, evaluated the effect of chronic consumption of “casabe” on the serum glucose concentration, weight, and length of the small intestine (SI). It was observed a reduction of glucose within 14 days of experiment, however, this level rose 16.3 % above the control group at 21 days (end time). The group fed with a diet containing casabe: “casaba Sucre” or “casabe Amazon” showed a significant increase in wet weight and length of the SI, the highest values 24.2 % and 8.94 % respectively was shown for the group “casaba Sucre”. The mechanism affected by the variation of glucose concentration, wet weight and length of the SI, is not yet known and still is in research. The varieties of “casaba Sucre”, “casaba Miranda”, and “casaba Amazon” were used to measure the effect on the serum concentration of triglyceride and total cholesterol. The group with the diet “casabe Amazon” showed the highest reduction value, 43% total cholesterol, 35% triglycerides; the variety “casaba Sucre” was 30-40% for total cholesterol and triglycerides. Also, the concentration of minerals showed a higher content of iron, copper and zinc compared with the control diet. The “casaba Sucre” showed the greatest iron and zinc content (6.49 and 36.76 ug/g diet respectively) and “casabe Amazon” showed the higher copper content 8.94 ug/g diet. A significant increase in the faecal excretion of iron and copper ($p<0.05$) was observed. The iron absorption increased 3.5% for “casaba Miranda”, 6% for “casaba Sucre” and 32% copper for “casabe Amazon”. The zinc absorption was less affected through the experimental time. The reduction of iron absorption for the “casabe Amazon” was associated to IDF, possibly due to the cationic interference process, suggested in the literature.

Keywords: Dietary Fiber, Casabe, Minerals, serum glucose, serum lipids

1. INTRODUCTION

Cassava (*Manihot esculenta Crantz*) is a tuber of the Family Euphorbiaceae of the genus Manihot, being the variety Euphorbiaceae *Manihot esculenta* which is more commonly known in South America. It is an edible root, very popular in many parts of Africa, Asia and South America. Other names for cassava are: yuca, casabe, and tapioca root.

Cassava is incredibly versatile; it can be boiled, baked, steamed, grilled, fried, crushed, or added to stews. It is often served with barbecue meat and many recipes include it.

Bitter grated cassava is used to make bread from cassava, which is the traditional crispy, flat, unleavened bread popular in Venezuela. There are two varieties of cassava; the sweet and the bitter (Scott et al., 2000). Both contain prussic acid (hydrocyanic acid) that can cause poisoning by cyanide, but roots of sweet variety of cassava have a very low cyanogenic potential (< 20.0 mg HCN per 100 dry weight). The cooking or the pressing of the root removes the poison. The Indians of Venezuela developed a method to extract the poison prussic acid from bitter cassava to make cassava bread "casabe". It involves peeling, washing, grating, and pressing using a “sebucan” (hanging sack). The pressing removes the poisonous liquid. Once separated from the juice, the pulp is dried in the sun and then made into bread or wrap (Farinha in Brazil and Garri in Nigeria). Known as “mañoco” in Venezuela, which is wrapped in banana leaves for storage. The process is laborious and whole villages would take part in its preparations. In Venezuela, for example, manioc root is food in used in many of the soups, "casabe" (flattened cassava cake), "naiboa" (cassava with brown sugar and grated white, forming a flat cake) and doughnuts (balls of dough of Yucca fries with brown sugar syrup).

Recent research has shown that eating a diet rich in dietary fiber may prevent or alleviate diseases such as cardiovascular diseases, diabetes, diverticulitis and cancer of the colon (Burkitt, 1971; Lairan et al., 2005; Kaline et al., 2007; Camparto et al., 2007; Queenan et al., 2007). Therefore, now a-days more fiber is being incorporated in the meals.

General Objective

To assess the effect of consumption of diets containing dietary fiber of the different varieties of Venezuelan “casabe” on the bioavailability of minerals, lipids, and proteins in an experimental model with rats.

2. Contents of Dietary Fiber

Cassava is part of the Venezuelan diet and it is valued by its calorific content, rich in dietary fiber (DF) and the mineral content. The consumption of cassava bread is not regular throughout the country, achieving high levels of consumption in the States of the eastern region of Venezuela: Bolivar, Monagas, Anzoátegui and Sucre and similarly in some States of the Centre of the country: Carabobo, Guárico, and Miranda; as well as in some indigenous peoples and communities of the Venezuelan Amazonas (Chonchol and Tovar, 1988). The cholesterol levels of these indigenous populations with a high cassava intake have been found to be lower than those from urban areas, and one reason for this could be the high content of dietary fiber (DF) in cassava (Rivera et al., 1993, Bosh and Camejo 1964).

A recent publication describes the contents of DF from the nutritional point of view, and according to the Venezuela food composition table, cassava has a 333 calories nutrient content per 100 g of food, 1.3 grams of protein, 0.6 grams of fat and 87.4 grams of carbohydrates, of which, 4.1 grams are DF; it also provides 3.1 milligrams of iron. (I.N.N, 1983)

Studies on the composition and nutritional characterization of cassava have indicated that it is a good source of DF, 5.67% insoluble dietary fiber(IDF); 4.92 % soluble dietary fiber(SDF). Other researchers have published a total DF content between a 3.40 to 3.78%. (Chonchol and Tovar, 1988; CIAT 2007). In Venezuela, some research has been done on the nutritional value and content of dietary fiber, resistant starches and other substances in cassava bread. Infante et al., (2013) report the content of DF of three varieties of "Casabe" from three different regions of the country using the method of Prosky et al., 1986. The "casabe" used in this study was bought in the local market three times per year in three locations in Venezuela: "Caripito" Monagas state; "Rio Chico" Miranda state and "Puerto Ayacucho" in the Amazonas state. As shown in Table 1, protein, fat, carbohydrates, ash and crude fiber as components of the cassava differs significantly ($P<0.05$) between each sample, which suggests differences between varieties of cassava plants used or the effect of soil on the plant. At present we have no botanical evidence to allow us to support one or more of these propositions. The moisture for the cassava samples ranged between 9.6 and 10.4% and the proteins between 1.1 and 1.3%. Although the fat content was below one percent, it showed a small range between 0.4 and 0.6%. The other components were ash between 1.4 and 1.7%, total carbohydrates between 86.9 and 87.3%, and crude fiber between 1.7 and 1.9% (Table 1).

Table 1. Proximal analysis of cassava bread in three different places of Venezuela (g/100g of food)

Locations	Moisture	Protein	Fat	Carbohydrates$^{\Delta}$	Ash	CF$^{\Diamond}$
Caripito	10.4± 0.02a	1.3± 0.08a	0.4 ±0.08a	86.9 ± 0.86a	1.4 ± 0.12a	1.8 ±0.22a
Rió Chico	9.9 ± 0.12b	1.1 ±0.17b	0.5± 0.12b	87.2 ±0.41b	1.5 ± 0.12b	1.9 ±0.08b
Puerto Ayacucho	9.6 ± 0.12c	1.2 ± 0.14c	0.6± 0.21c	87.3 ± 0.29c	1.7 ±0.21c	1.7 ±0.09c

Values are means of at least 4 independent determinations ± standard deviation.
◊ CF: Crude fiber.
Δ: The carbohydrates were obtained by difference.
Means on the same column with different superscript are significantly ($P<0.05$) different.

Table 2. Dietary fiber and pectin contents in cassava bread in diverses locations in Venezuela (g/100 g of food)

Locations	Insoluble Dietary Fiber	Soluble Dietary Fiber	Total* Dietary Fiber	Pectin
Caripito	4.6 ± 0.08a	0.8 ± 0.08a	5.4	0.34 ± 0.08a
Rio Chico	4.5 ± 0.22b	0.7 ±0.12b	5.2	0.31 ± 0.08b
Puerto Ayacucho	4.7 ± 0.12c	1.6 ± 0.17c	6.3	0.61± 0.07c

Values are means of at least 4 independent determinations ± standard deviation.
(*): Total Dietary fiber by sum (IDF + SDF).
Source: Infante et al., 2013.

Table 2 shows the comparison of dietary fiber between the three varieties of "casabe". The results indicate significant differences ($P<0.05$) for each variable when compared between different samples. The total dietary fiber, ranged between 5.2 and 6.3%, and the insoluble dietary fiber content was similar for all three samples, averaging 4.6%. While the soluble fraction of dietary fiber showed a range between 0.7 and 1.6%, being the highest value of SDF for the cassava "Puerto Ayacucho".

Some other studies by Chonchol and Tovar, 1988, reported a value of IDF between 4.92 and 5.67%, SDF values between 3.4 and 3.78% and a starch content that ranged from 83.4 to 88.7% of "casabe" obtained from different places of Venezuela. These values were very close to those described by Infante et al., (2013). However Chonchol and Tovar used a different enzymatic method for measuring DF (Asp et al., (1983)), which is an enzymatic-gravimetric method known to retain starch in the dietary fiber fractions, which could explain the significant differences between their SDF higher values and the shown in Table 2.

Table 3. Dietary fiber and starch from cassava (%)

	Mean	SD	C.V
NDF	4.6	50.45	9.84
ADF	2.79	0.43	15.48
Hemicelluloses	1.96	0.25	13.07
Cellulose	1.76	0.29	17.37
Lignin	0.56	0.04	7.34
Pectin	1.17	0.10	8.74
Starch	83.02	5.26	6.34

Values are in relative percentage for dry samples.
NDF estimates cellulose, lignin and hemicelluloses.
ADF estimates cellulose and lignin.
Abbreviations: NDF: neutral detergent fiber; ADF: acid detergent fiber; C.V.: coefficient of variation.
Source: Rivera et al., 1993.

Rivera et al., 1993, using "casabe" from Rio Chico "San Jose", Estado Miranda, Venezuela, made the determination of dietary fiber using neutral detergent fiber (NDF) method modified by Holechek and Vavra to reduce the time of the assays, as well as the quantities of sample and reagents used which allows processing of a larger number of samples. This method was applied to samples of "casabe" previously digested by pancreatic enzymes. NDF is a method that allows the estimation of cellulose, lignin and hemicelluloses while acid method detergent fiber (ADF) estimated cellulose and lignin. However, this method of NDF Southgate et al., 1978, is still used for the determination of insoluble in legumes and cereal fiber (Candela et al., 1997). As can be seen in Table 3, the IDF and SDF dietary fiber values were close to those described by Infante et al., 2013 and Chonchol and Tovar, 1988, but were lower than those reported by Moron et al., 2013.

PHYSIOLOGICAL ASSAYS USING "CASABE"

An assessment of the rate of digestibility (DI) of "casabe" was performed by Chonchol et al., 1988. Briefly, after three hours of washing with ethanol, cassava flour was incubated with porcine pancreatin (1 mg/ml) as a source of amylase, and the external environment of the dialysis bag samples for the determination of sugar by reducing methods were taken. After 2 minutes of incubation a value close to 20% of hydrolysis was observed. This increase of the amylases sensitivity may be the consequence of the process of elaboration of the cassava, notably the method of extending the cassava flour over the hot surface of a "budare". The results described above are in accordance with high human response of glycemic to "cassava" found by Melania et al., 2002. They determined the Glycemic Index(GI) and insulin index (II) to a group of volunteers who consumed a diet with boiled food cassava or "casabe" as processed cassava, standard techniques were used to measure plasma glucose, serum or plasma insulin, for 0, 15, 30, 45, 60, 90 and 120 minutes. After the ingestion of cassava or "casabe" equivalent to 50 grams of glucose consumption.; "casabe" or cassava produced the higher IG with values above the pattern of glucose values, 117, 55±33, 81 and 108, 19±20, 60, respectively.

The author concludes that the carbohydrate content regarded as complex (polysaccharides), or simple carbohydrates (mono and disaccharides) of a food, cannot be used to predict the effects of blood insulin and glycemic response.

EFFECT OF "CASABE" ON THE LIPID PROFILE

For the realization of this work, 20 rats males rat of the *Sprague Dawley* strain were used, with a starting weight of 150g which were divided randomly into four groups of 5 rats each(Moron, 2008).

Table 4. Percentage Composition of the Experimental Diets (g/100 g of the mix) a

Ingredient	Control	Cassava Miranda	Cassava Sucre	Cassava Amazonas
ISOLATED SOY BEAN	18	18	18	18
CORN OIL	8	8	8	8
CORN STARCH	68.5	--	--	--
CASSAVA BREAD	--	68.5	68.5	68.5
MIXTURE OF VITAMINS	1	1	1	1
MIXTURE OF MINERALS 4	4	4	4	4
BITARTRATE OF COLINE	0.2	0.2	0.2	0.2
L- METIONINE	0.3	0.3	0.3	0.3

a. Mix AIN-76 of Harlan Teklad.
Source: Moron 2008.

Table 5. Growth, food consumption and efficiency of diets for different treatments

	Experimental groups Control	"Casabe" Miranda	"Casabe" Sucre	"Casabe" Amazonas
Initial weight (g)	151.26 ± 20.78[a]	141.51 ± 18.54[a]	139.15 ± 8.76[a]	145.79 ± 19.16[a]
Weight gain (g / 21d)	99.81 ± 11.18 [a]	122.60 ±11.39[b]	109.29 ± 11.33[b]	127.88 ± 16.41[b]
Food consumption (g/21d)	279.33 ± 6.36[a]	315.27 ± 7.54[b]	310.39 ± 7.25[b]	301.44 ± 7.97[b]
Diet efficiency (%)	36.40 ± 5.90[a]	38.40 ± 1.52[a]	36.80 ± 5.02[a]	40.75 ± 1.26[a]

Note: The table shows Mean and deviation standard of 5 animals. The values in the same row with different letters are statistically significant, applying parametric ANOVA test of two factors ($p < 0,05$).
Source: Moron 2008.

Table 5 shows three experimental groups that differed from the control because the corn starch was replaced by a specific variety of cassava with the same percentage. Cassava-fed groups were identified according to the origin of the "casabe", the first group was controlled without cassava in their basic diet, the second group was known as "casabe Miranda", the third group designated "casabe Sucre", and the fourth group identified as "casabe Amazon". The same samples of cassava described by Infante et al., 2013 were used in this study. Starch content of the diet was replaced by "casabe" grinding and sieving and by considering that a diet that contributes 18% of protein meets the requirements of the rat (American Institute of Nutrition (1977)). Both the food and water were offered *ad libitum* during the entire experimental period. The rats were subjected to a break-in period for three days, on the third day they received, water and food for rodents (Protinal®) *ad libitum*. At the end of the chronic period of intake of their respective diets, the animals were anesthetized using Sodium Pentothal and then an incision in the chest cavity was made to obtain blood sample by cardiac puncture. Subsequently, blood samples were centrifuged at 1500 rpm for separating plasma from red blood cells and then frozen at - 20 ° C until further analysis.

The effect of consumption of different varieties of "casabe" on the relative weight gain, and index of efficiency for food (FER) in rats, are shown in Table 5. All the animals started the experimental period of 21 days with the same conditions of body weight. At 21 days it was observed that rats fed with diets supplemented with "casabe", showed a relative body weight gain, significantly higher, compared to animals fed a diet without "casabe" (control), differing in 28% the group fed with cassava Amazon, followed by the group fed with cassava Miranda with 23%. The efficiency index food (FER), determines the degree of efficiency of a diet as a source of energy for an organism. In this study, no significant differences were found between controls fed without "casabe" and "casabe"-fed groups, suggesting that groups fed with "casabe" consumed greater amounts of fiber than the control group, which did not cause reduction in the bioavailability of nutrients as to affect the growth of the rats per gram of diet consumed.

Table 6, shows the effect of the inclusion of "casabe" in the experimental diets on serum total cholesterol. A significant decrease in the concentration of serum cholesterol was observed in all groups fed with this source of DF ($p < 0,05$). In addition the group "casabe Amazon" showed the greatest reduction in the concentration of total cholesterol: 45% compared to controls fed without cassava, followed by the group "casaba Miranda" 38% and group "casaba Sucre" 32% respectively. Not significant differences between varieties of "casabe" Sucre and Miranda were observed. Soluble dietary fibers have been proposed to produce a delay in the absorption of nutrients in the absorption of nutrients; therefore the author suggests that the lower absorption of lipids in the gastrointestinal system could be associated with the increase of viscosity produced by this type of fiber, which would lead to the formation of a viscous matrix gel that slows down the diffusion of nutrients from the light to the intestinal surface and this could explain the greater reduction of serum total cholesterol concentration observed in the group "casabe Amazon".

Table 6. Rats fed diets with different varieties of casabe. Serum concentration of total cholesterol: mg/dl

Cassava varieties	Mean	Deviation standard
Control Group	110	14c
Group "casabe" Miranda	72	6ab
Group "casabe" Sucre	78	6b
Group "casabe" Amazonas	60	6a

Values in the same row with different letters are statistically significant, using the parametric ANOVA test of two factors ($p < 0, 05$).

Source: Modified from Moron 2008/

Table 7. Rats fed diets with different varieties of casabe. Serum concentration of cholesterol-hdl: mg/dl

Cassava varieties	Mean	Deviation standard
Control Group	22.3	5a
Group "casabe" Miranda	23.07	3a
Group "casabe" Sucre	28.46	6.92a
Group "casabe" Amazonas	28.84	9.23a

Values in the same row with different letters are statistically significant, using the parametric ANOVA test of two factors ($p < 0,05$).

Source: Modified from Moron 2008.

Table 7 show a comparison between the control group and the groups of rats fed with "casabe" diets. The groups "casabe" Sucre and Amazon showed an increase in the concentration of HDL-cholesterol 29.5% and 31.52% respectively, but without statistical significance.

Studies with hamster, using non-digestible polysaccharides, extruded cassava starch (E-CS), extruded cassava resistant starch (E-CS-RS) and extruded cassava starch with added oat (E-CS-RS-AO), as a source of fiber; shown that HDL-cholesterol, was higher in the group fed diet (E-CS-RS-AO) (195 mg/dl) and was considerably different from the group fed diet (E-CS-RS) (158 mg/dl) and (E-CS) (171 mg/dl) respectively. These same groups presented values of total cholesterol in serum considerably lower, when compared with control group. Current information on the effect of dietary fiber on the levels of HDL-cholesterol is contradictory, so there are reports where the intake of isolated dietary fiber or foods sources of dietary fiber produced a reduction in the levels of HDL-cholesterol (Rosa et al., 1988). Other studies reported increased levels of HDL (Dabai et al., 1996), and some other studies report that there are no effects on HDL-Cholesterol levels. (Prasad and Stuglin, 2005).

Table 8. Rats fed diets with different varieties of casabe. Serum concentration of LDL-cholesterol: mg/dl

Cassava varieties	Mean	Deviation standard
Control Group	54.73	10c
Group “casabe” Miranda	20.52	10.52a
Group “casabe” Sucre	28.42	12.63b
Group “casabe” Amazonas	10.52	6.31a

Values in the same row with different letters are statistically significant, using the parametric ANOVA test of two factors (p < 0,05).

Source: Modified from Moron 2008.

Table 8 shows a significant decrease in the fraction of LDL-Cholesterol in all groups of animals fed with “casabe “as a source of DF. The largest decrease occurred in the group "casabe Amazon”, 81% in comparison with control group, followed by the groups "casabe Miranda” and Sucre who showed a reduction of 62.7% and 48.3% respectively

Finally the serum triglyceride concentration shown by these groups fed with diets including “casabe” are shown in Table 9. As can be seen there is no effect on the concentration of serum triglycerides. This could be due to the high content of starch present in the “casabe”, (81 %) which is comparable to the value found in the control group.

These results are consistent with those reported by Rosa and Costa, 1998, who found no significant difference on the total triglycerides in serum of rats fed diets supplemented with insoluble dietary fiber of *Phaseolus vulgaris* (black beans, “carioquiña”, and red) at a rate of 30% on the dry basis, for a period of 21 days, compared to the control group that was not fed with dietary fiber.

Table 9. Rats fed diets with different varieties of casabe. Serum concentration of triglicerids: mg/dl

Cassava varieties	Mean	Deviation standard
Control Group	141.4	14.28a
Group “casabe” Miranda	137.14	5.71a
Group “casabe” Sucre	135.7	7.14a
Group “casabe” Amazonas	135.7	8.57a

Values in the same row with different letters are statistically significant, using the parametric ANOVA test of two factors (p < 0,05).
Source: Modified from Moron 2008.

The effects of cassava in the lipid profile are the important part of this foodstuff obtained from cassava flour (*Manihot esculenta*) as a future food of large benefits, allows storage for a long time, easy handling. It can accompany other foods during feeding, etc. Adding that the tuber from which flour is obtained is harvested in large quantities in tropical countries.

Effect of “Casabe”, on Weight and Length of the Small Intestine of Rats

Table 10 presents the measures of length and weight of the intestines of rats fed with diet control and diet with different varieties of “casabe”, as a source of DF of the same experimental trial described above, expressed in total grams, grams / 100g body weight, and g/100 g of food.

From these data can be seen that both, the weight of intestines and its length showed a significant increase ($p < 0.05$) in groups of rats fed diets with different varieties of “casabe”, as source of DF, compared with the control group. The comparison of these measures, demonstrated that the inclusion of cassava in the diets produced intestines longer and heavier, than rats fed with diet control. This effect was independent of the variety of used “casabe” (Redondo et al., 2004).

However it should be noted that if the weight of the intestine was corrected by final body weight and by grams of food consumed, there were no differences between group control and group fed with “casabe” diets, suggesting that the intestine size was suitable for the weight and the volume of consumed food, independent that the latter was considerably higher in "casabe" groups.

Tabla 10. Variation of weight and length of the small intestine in rats fed diets with different varieties of casabe

	Experimental groups*			
Small Intestine	**Control**	**Casaba Miranda**	**Casaba Sucre**	**Casaba Amazonas**
Length (cm)	109. 3 ± 4.57a	124.6 ± 8.38 b	120.7 ± 4.92 b	123.6 ± 8.32 b
Wet weight (g)	4.50 ± 0.78	5.59 ± 0.76 b	5.12 ±0.33 b	5.38 ± 0.76 b
Weight (g/100 weight)	2. 27 ± 0.15 a	2.12 ±0.15 a	2.05 ± 0.19 a	2.16 ±0.16 ª
Body weight (g/100 g food)	2. 50 ± 0.26 a	2.53 ±0.18 a	2.56 ± 0.10 a	2.55 ± 0.16 a

* The table shows the Mean and deviation standard of 5 animals. The values in the same row with different letters are statistically significant, applying parametric ANOVA test of two factors (p < 0, 05).

Source: Moron 2008.

The results found in this study coincide with observations of Elsenhans et al., 1981, who found an increase in the length of the intestine in response to the consumption of non-digestible carbohydrates. These authors attributed these changes to the mechanical stimulus exercised by the volume and peristaltic flow of the intestinal mucosa and adjoining muscle layer. Other researchers have also warned that the DF consumption increases the weight and thickness of the intestinal mucosa (Ecknauer et al., 1981), as well as the replacement of cells, the life time and the migration of the enterocytes from along the villus-crypt axis, making the villi finer and tighter microvilli (Tetsuguchi, 1997).

Mineral Content of Cassava

DF or some of its components can alter the absorption of carbohydrates, proteins, lipids and cations by a process of dilution of the fiber on the food bowl or intestine morphological alterations and effects on intestinal mucosa (Zhang et al., 2013). Reducing the bioavailability of certain minerals and electrolytes due to its absorption and elimination by faeces, is one of the adverse effects traditionally attributed to dietary fiber (Falcón et al., 2011)

The mechanisms by which the DF intervenes in the process of digestion-absorption in the small intestine is not clear, which raises the need for fundamental studies that relate the physicochemical properties of the sources of fiber and its compounds. The purpose of the study cited in this chapter, was to determine the effect of the consumption of different varieties of Venezuelan “casabe” on the absorption of minerals in rats as experimental model. To comply with said purpose Moron et al., 2013, designed a biological assay, briefly: 4 types of diets were developed, a diet control without DF, following the recommendations of the American Institute of Nutrition 1977, with a contribution 18% isoprotein, which meets the needs of the rats. Three experimental diets that differed from the control, by the substitution of corn starch by a specific variety of cassava: "casabe Miranda", "casabe Sucre" "and casabe Amazon". The composition of these diets is described in Table 4.

20 male rats *Sprague Dawley*, with an average initial weight of 150g, were divided randomly into four groups of 5 rats each. The rats were fed *ad libitum* during the whole trial (21 days) with the above diets. The management of the rats was made according to international standards for animal testing. The weights of animals and food intake were recorded inter-daily.

Content of Iron, Copper and Zinc in Experimental Diets

Table 11 shows the values of contents of, iron, copper and zinc in experimental diets. As observed the addition of “casabe” diets had a higher content of iron, copper, and zinc, compared with diet control. The “casabe Sucre” diet presented the highest iron and zinc content, while the “casabe Amazon” diet showed the highest contents of copper.

Table 12 shows the values of the fecal excretion of iron, copper and zinc in rats fed diets with and without cassava. There is a significant increase in the excretion of iron and copper in all groups of rats fed with “casabe” diets. The “casaba Sucre” diet showed an increased fecal excretion of copper, 77.7% in comparison with the "control diet", followed by the "casabe Miranda" diet, 38.8% and finally the "casabe Amazon" diet 16.7%. This order does not show differences statistics.

Table 11. Content of iron, copper, and zinc in experimental diets

Diets	Iron µg/g	Copper µg/g	Zinc µg/g
Control	4.64	6.92	34.16
"Casabe" Miranda	6.21	7.70	36.68
"Casabe" Sucre	6.49	7.77	36.76
"Casabe" Amazonas	5.00	8.94	36.45

Source: Moron et al., 2013.

Table 12. Fecal excretion of copper, iron and zinc in rats fed diets with and without addition of casabe (MG of copper, iron, ZINC/total in feces) 21 days after the experimental period

GROUPS	COBRE	HIERRO	ZINC
Control	0.18 ±0.03a	0.21 ±0.04a	0.06 ±0.02ab
"Casabe" Miranda	0.25 ±0.03b	0.41 ±0.06 b	0.09 ±0.01b
"casabe" Sucre	0.32 ±0.01c	0.37 ±0.04b	0.07 ±0.01ab
"Casabe" Amazonas	0.21 ±0.06ab	0.49 ±0.19 b	0.05 ±0.03a

The table shows the Mean and deviation standard of 5 animals. The values in the same row with different letters are statistically significant, by applying one-way ANOVA test ($p < 0.05$).

Source: Modified from Moron et al., 2013.

Faecal excretion of iron showed the higher values compared to the rest of Cu and Zn minerals; the "casabe Amazon" diet showed the highest value of excretion 133.3% followed by the "casabe Miranda" diet with 95.2% and finally the "casabe Sucre" diet with 76.1%.

Relationship with fecal excretion of zinc, found that the addition of "casabe" diets, did not produce any effect on fecal excretion between 7 and 14 days of the experiment. However, at 21 days it was found that the "casabe Miranda" diet, showed a significant increase in fecal excretion of zinc approximately 50% in comparison with the control diet, followed by the "casabe Sucre" diet with 16.6%, while group "casaba Amazon" diet showed a decrease of 16.7% compared to control diet.

Excretion is particularly noticeable for Cu and Fe; it may also consider the contribution of "casabe" in mineral contents as shown in Table 11, where the content of both minerals was higher than the control group.

Effect of "casabe diets" on the absorption of copper and iron in rats during the experimental period.

Table 13 shows the percentage of absorption of copper in rats fed with control diet and with "casabe" diets at 21 days of trial. The authors of this study made three shots of samples every seven days. For the 7th experimentation day, "casabe" Sucre and Miranda showed a decrease in the absorption of copper but without significant Statistical impact, in fact this trend is reduced on days 14 and 21.

Table 13. Effect of diets with different varieties of Venezuelan "Casabe" on the absorption of copper in rats at 21 Experimental Days

DIET	%ABSORPTION OF COPPER	STANDARD DEVIATION
CONTROL	40.52	6.14ª
"Casabe" SUCRE	42.98	6.1ª
"Casabe" MIRANDA	45.43	6.14ab
"Casabe" AMAZONAS	52.8	5.89b

The table shows the Mean and deviation standard of 5 animals. The values in the same column with different letters are statistically significant, by applying one-way ANOVA test, followed by the Duncan test ($p < 0,05$).

Source: Modified from Moron et al., 2013.

The interesting point regarding these values is that none of the breakpoints of the "casabe Amazon" diet showed inhibition of absorption of copper, and always remained above the value of the control diet with statistical significance. This trend was continued during trial, and also the two remaining groups presented an increase in the absorption of copper, with statistical significance ($p<0.05$).

Table 14. Effect of diets with different varieties of Venezuelan "casabe" on the absorption of iron in rats at 21 days

DIET	% ABSORPTION OF IRON	STANDARD DEVIATION
CONTROL	89.36	2.12b
"CASABE" SUCRE	85.1	4.2ab
"CASABE" MIRANDA	85.1	2.1ab
"CASABE" AMAZONAS	78.72	8.51a

The table shows the Mean and deviation standard of 5 animals. The values in the same column with different letters are statistically significant, by applying one-way ANOVA test, followed by the Duncan test ($p < 0.05$).
Source: Modified from Moron et al., 2013.

In terms of absorption of copper, it can be seen that "casabe Miranda and Sucre" diets showed a decrease in the absorption of copper, but only in the initial stage of the assay (day 7). While the "casabe Amazon" diet, showed an increase in the absorption of copper during all the essay period associated with a decrease in faecal excretion of this mineral.

It should be noted that once the content of minerals was quantified in the experiment, the higher bioavailability of this mineral during the experimental period, coupled with a decrease in faecal excretion of copper was equally evident.

On the interaction of minerals with DF, soluble and insoluble fibers behave unevenly in different segments of the small intestine.

It has been pointed out that most of the minerals are absorbed in the small intestine, but some such as copper, may be absorbed partially in the stomach and now arises that the colon is a place where also this mineral can be absorbed which would explain why the inhibition of the mineral absorption is not so pronounced. (Frohlich, 1995) These results could indicate that this variety of cassava does not show an association between the content of DF, particularly insoluble, on the absorption of copper.

The percentage of absorption of iron was analyzed for the different experimental groups; it was found that only “casabe Amazon” diet was adversely affected during the experimental period.

Increased absorption of iron in “casabe Miranda and Sucre” groups could mean a higher intake of this mineral due to these “casabe varieties” in the diet. However, several researchers have observed that the availability of iron could be reduced by the presence of FD (Bueno et al., 2012; Salas and Garcia, 2000). It has been described that a human being is able to adapt himself to relatively high amounts of DF, balancing the inputs and outputs of these minerals within a few weeks. (Frohlich, 1995).

The interference of DF with the process of absorption of cations has been controversial. It has been suggested that a high consumption of DF in adulthood (26 g/day) does not affect significantly the food balance of iron, calcium, zinc, and copper. Today many authors consider that the cation exchange capacity of the dietary fiber, in normal dietary components, is not capable of producing disorders in the absorption of minerals. (Sandstrom et al., 2000).

To explain the little negative effect of DF on the uptake of cations, it has been suggested that ions together with acidic polysaccharides are released to reach the large intestine, since these carbohydrates are susceptible to degradation by colonic bacteria. (Coudray et al., 2005). Of special interest is the study analyzed in this book chapter, the variety of "casabe Amazon", which can be associated with the higher content of DF, in particular with the insoluble dietary fiber, negatively affected the absorption of iron and it could be attributed to the process of interference cationic described above.

It becomes relevant to point out that even when all the experimental groups were found to have an increase in faecal excretion of iron, the amount consumed in the diet allowed a better absorption and utilization of the same.

Zinc, was the mineral least affected in their absorption throughout the entire experimental time, highlighting a positive effect on the absorption of zinc at the end of the essay in the group “casabe Amazon” diet; the zinc content was similar in the three “casabe varieties” groups. An association between the content of soluble and insoluble, FD on the absorption of zinc was found.

While in Venezuela precooked corn flours for human consumption, are enriched with iron and vitamins (FUNDACREDESA, 2002) and for the time being it has significantly improved the nutritional status of the child population. Equally certain is that the severe anemia and iron deficiency continues to be a clinical manifestation in these populations (Baron et al.,

2007). Similarly the copper and zinc are associated with physical and mental growth of the child population (Cordero et al., 2010.). To include a complementary food as "casabe" in the children diet, considering it contains significant amounts of iron, copper and zinc, seems an apparent solution at the problems of anemia and immune deficiencies in child population.

The production of industrial cassava flour develops in Venezuela at low level. To date the harvest of cassava plant is insufficient, even though it represents an enormous potential for both human and animal consumption (FAO/FIDA, 2000). Petroleos de Venezuela (PDVSA) aims to increase the cultivation of cassava to 75,000 hectares, because in addition to the aim of using them for human consumption, animal and industrial use, studies are performed to obtain alcohol from the cassava (PDVSA, 2009). On the other hand, there is a program of improvement of cassava with the purpose of obtaining improved clones containing low contents of hydrocyanic acid. These and other similar studies are in development at the Institute of agricultural research (INIA) in Venezuela (Fuenmayor et al., 2012).

Major development work on this crop has taken place in the artificial drying of cassava root to produce flour for the food Industry. Cassava flour is produced using different forms of thermal processing and its consequent characterization of its chemical composition, physical and functional properties. Perez et al., 2007, evaluated the physical, chemical and rheological properties of heat-treated cassava flours. They concluded that thermal pregelatinization methods applied in the study of cassava flour, seemed to cause certain changes in their physical and chemical and rheological properties. In particular they showed a reduction in the tendency to retrograde, a decrease in the consistency and rate of absorption, increase in output, stability, tolerance and time of rupture, which then recommended it as a potential ingredient in conventional products and the development of new products. (Montagne et al., 2009, Perez et al., 2007). In short, increasing the production of cassava with a low content of acid hydrocyanic acid, improving the industrial production of cassava, and incorporating it into certain products formulation such as, baked goods could lead to its consumption in populations of children in different forms of dietary supplement. It should be a proposal to be considered for the improvement of the nutritional status of the population.

ACKNOWLEDGMENTS

This research work was supported by grant Nº PI-09-8229-2011/1. Consejo de Desarrollo Científico y Humanístico, Universidad Central de Venezuela (CDCH-UCV).

We acknowledge to Mr. Stefan Regli for the valuable reading of this manuscript.

REFERENCES

American Institute of Nutrition. Report of the American Institute of Nutrition *ad hoc* committee of standards for nutritional studies. 1977. *J. Nutr.* 107:1340-1348.

Asp, NG, Johansson CG, Hallmer H,Siljestrom M. Rapid enzymatic assay of insoluble and soluble dietary fiber. *J. Agric. Food Chem.* 1983. *31*: 476-482.

Baron MA, Solano L, Paez MC, Pabon M. Estado nutricional de hierro y parasitosis intestinal en niños de Valencia, Estado Carabobo, Venezuela. *An Venez Nutr*. 2007. 20: 5-11.

Bosch, V. and Camejo, G. Serum lipoprotein of Amazonian indians and inhabitants of an urban area of Venezuela fractionated by preparative ultracentrifugation. *Metabolism*. 1964.133, 11: 456-461.

Bueno L, Marchini J, Dutra J. Cálcio, goma guar parcialmente hidrolisada e triacilglicerídeos de cadeia média interferentes na disponibilidade do ferro não-heme em soluções nutrientes. *Rev. Bras. Nutr. Clin.* 2012. 27: 213-217.

Burkitt DP. The epidemiology of cancer of the colon and rectum. *Cancer* 1971. 28: 3 -13.

Camparto et al.,; Camparto G, Pilotto A, Franze A, Franceschi M, Di Mario F. (2007). Diverticular diesease in the elderly. *Digestive Disease.* 2007. *25*:151 -159.

Candela M, Astiasaran I, Bello J. Cooking and warm-holding: effect on general composition and amino acids of kidney beans (*Phaseolus vulgaris*), chickpeas (*Cicer arietinum*), and lentils (*Lens culinaris*). *J. Agric. Food Chem.* 1997, 45, 4763−4767.

Chonchol N, Tovar J. Dietary fiber content and starch digestibility in cassava bread. *Nutr. Reports Intern.* 1988. *38*:437- 443.

CIAT. 2007. CIAT Cassava: A crop for hard times and modern times. In focus crop commitments. Accessed October 2007 at www.ciat.cgior.org

Cordero M Raimundo E, Ramón Benito Infante, Rita Chacón, Patricio Hevia. Cinc y cobre plasmático. Su relación con el estado nutricional en niños menores de siete años. *Revista de la Facultad de Medicina.* 2010. Volumen 33: 22-28.

Coudray C, Rambeau M, Feillet Ch, Tressol J, Demingue C, Gueux E. Dietary inulin intake and age can significantly afftect intestinal absorption of Calcium and magnesium in rats: a stable isotope approach. *Nutrition Journal.*2005. 4:29-34

Dabai F, Walker A, Sambrook I, et al., Comparative effects on blood lipids and faecal steroids of five legume species incorporated into a semi-purified hypercholesterolaemic rat diet. *Br. J. Nutr.* 1996. 75: 557-571.

Ecknauer R., Sircar B., Johnson, L. Effect of dietary bulk on small intestinal morphology and cell renewal in the rat. *Gastroenterology.* 1981. 81: 781 – 786.

Elsehans B., Blume R.& Caspary, W. Long term feeding of unavailable carbohydrate gelling agents: Influence of dietary concentration and microbiological degradationon adaptative reponses in the rat. *Am. J. Clin. Nutr.* 1981. 34: 1837 – 1848.

Falcón M, Barrón J, Romero A, Domínguez M. Efecto adverso en la calidad proteica de los alimentos de dietas con alto contenido de fibra dietaria. *Rev. chil. nutr.* 2011. 38:356-67.

FAO/FIDA. Fondo internacional de desarrollo agrícola/Organización de las Naciones Unidas para la Agricultura y la Alimentación. La economía mundial de la yuca: hechos, tendencias y perspectivas. Roma, Italia. 2000. 59 p.

Frohlich W. Bioavailability of micronutrients in a fibre-rich diet especially related to mineral. *Eur. J. Nutr.* 1995. 40:S116-22.

Fuenmayor F, Montilla J, Albarran J, Perez M, Vaccarino L, Segovia V. Evaluación y selección de clones de yuca (*Manihot esculenta* Crantz) del Plan Nacional de Semilla del INIA-Venezuela. Resultados preliminares. *Rev. Cient UDO Agricola.*2012. 12: 17-24.

FUNDACREDESA. Impacto poblacional en Venezuela por el enriquecimiento con hierro y vitaminas de las harinas pre cocidas de consumo humano. Ministerio de Salud y Desarrollo Social/UNICEF.FUNDACREDESA. Una visión integral de Venezuela XXV años. Primera Edición. *Caracas,* 2002.

Holechek, J.L. and Vavra, M. Comparison of micro and macro digestión methods for fiber analysis, *J. Rug. Mgmt.* 2012. 35:799-801.

I.N.N. Tabla de composición de alimentos para uso práctico M.S.A.S. Publication N° 40, Caracas-Venezuela. 1983.

Infante Benito, Garcia Omar, Rivera Carlos. Characterization of dietary fiber and pectin of cassava bread obtained from different regions of Venezuela. *Rev Chil Nutr.* 2013. 40: 169-173.

Izquierdo, M. Índices Glucémicos e Insulinémicos de Alimentos Venezolanos. UCV. Tesis Título Doctorado en Ciencias, Mención: *Nutrición Clinica.* 2002. 1 – 68.

Jenkins DJ, Marchie A, Augustin LS, Ros E, Kendall C. Viscous dietary fibre and metabolic effects. *Clin Nutr.* 2004. 1: S39-49.

Jenkins, D., Wloever, T. Kockaday T., Leeds A., Howarth R., Bacon S., Apling, E. & Dilawari J. Treatment of diabetes with guar gum: reduction of urinary glucose loss in diabetics. *Lancet.* 1977. 2: 779 – 780.

Kaline K, Bornstein SR, Bergmann A, Hauner H, Schwarrz H. The importance and effect of dietary fiber in diabetes prevention with particular consideration of whole grain products. *Horm Metab. Res.* 2007. *38*: 687 – 693.

Lairan D, Arnault N, Bertrais S, Planells R, Clero E, Hercherg S, Boutrom-Rualult MC. Dietary fiber intake and risk factors for cardiovascular disease in French adults. *Am. J. Clin. Nutr.* 2005. *82*:1185 -1194.

Montagne JA, CR Davis, SA Tanumihardjo. Nutritional value of cassava for use as a staple food and recent advances for improvement.. *Comprehensive Reviews in Food Science and Food Safety.* 2009. 8: 181-194.

Morón C Mirla C; Ana V Ávila A; Pablo I Hernández R. Efecto del consumo de casabe venezolano sobre la absorción de minerales en un modelo experimental con ratas. *INHRR.* 2013. In press.

Morón Mirla. 2008. Efecto del consumo de fibra dietética proveniente de diferentes variedades de casabe venezolano sobre la absorción de algunos nutrientes en un modelo experimental con ratas. Thesis work for promotion to profesor agregado, Universidad Central de Venezuela.

Pérez EE, Amaiz ML, González Z, Tovar J. Production and characterization of cassava (*Manihot esculenta* Crantz) flours using different thermal treatments. INCI, 2007. 32: 1-8.

Petróleos de Venezuela, S.A. (PDVSA). 2009. Proyecto Agro energético etanol- combustible. Disponible en: http://www.pdvsa.com/siembra_

refinacion_internet/pdf/copia_de_etanol.pdf. Última visita 20 de marzo de 2012.

Prasad K, Stuglin C. Effect of flaxseed consumption on blood pressure, serum lipids, hemopoietic system and li ver and kidney enzymes in healthy humans. *J. Cardiovasc. Pharmacol. Ther.* 2005; 10: 7-23.

Prosky L.; Asp. N-G; Furdal.; Devries, J.G.; Schweizer, T.P.F. and Harland, B.F. Determination of total dietary fiber in food products: Collaborative study. *J. Assoc. Offic. Anal. Chem.*1986. 68: 677-679.

Queenan K, Stewart M, Smith K, Thomas W, Fulcher R, Slavin J. Concentrated oat beta-glucam, a fermentable fiber, lowers serum cholesterol in hypercholesterolemic adults in a randomized controlled trial. *Nutrition Journal.* 2007; 6:6 doi:10.1186/1475-2891-6-6 http://www.nutritionj.com/content/6/1/6.

Redondo Bonive, Mendez J, Abraham A, Moron M, Infante B. Study of chronic cassava bread consumption on the serum glucose level and the weight and length of the large intestine in rats. *Acta Cient Ven* 2004. 55.(sup 1): 199.

Rivera C, Gerardi A, Infante R, Carrasco H, Rodriguez O. Dietary fiber analysis of cassava using gravimetric methods. *Arch. Lat. Nutr.* 1993. 43: 73 - 74.

Rosa C, Costa N, Leal P, Oliveira T. Efeito do feijão (*Pha seolus vulgaris,*L) sem casca na redução do colesterol san güíneo de ratos hipercolesterolêmicos. *Arch. Lat. Nutr.* 1998; 48:299-305.

Salas J, García P. Dieta rica en fibra. En: Nutrición y dietética clínica. Barcelona: Ediciones Doyma. Barcelona, España. 2000.

Sandstrom B, Bugel S, McGraw B, Price J, Reid M. A high oat- bran intake does not impair zinc absorption in humans when added to a low-fiber animal protein-based diet. *J. Nutr.* 2000. 130: 594-599.

Scott GJ; Rosegrant MW; Ringler C. Root and tubers for the 21st Century. Trends, projections, and policy options. International food policy research institute (IFPRI)/CIP. Centro Internacional de la Papa. Washington, USA. 2000. 64 p.

Southgate D. Dietary fiber analysis and food sources. *Am. J. Clin. Nutr.* 1978. 31: 107S-110S.

Tetsuguchi M, Nomura S, Katayama M, Sugawa Y. Effects of curdlan and gellan gum on the surface structure of intestinal mucosa in rats. *J. Nutr. Sci. Vitaminol.* 1997. 43: 515-527.

Zhang W, Li D, Liu L, Zang J, Duan Q, Yang W, et al., The effects of dietary fiber level on nutrient digestibility in growing pigs. *J. Anim. Sci. Biotechnol.* 2013. 4:17-22.

In: Cassava
Editor: Francis P. Molinari
ISBN: 978-1-63321-031-8

Chapter 6

VISCOSITY, AGGLOMERATION BEHAVIOR AND TABLETING CHARACTERISTICS OF CASSAVA STARCH MODULATED BY WET GRANULATION: A COMPARATIVE STUDY

John Rojas, Marcel Yepes and Yhors Ciro
Department of Pharmacy, School of Pharmaceutical Chemistry,
University of Antioquia, Antioquia, Colombia

ABSTRACT

Starch is one the most common natural polymers used for the preparation of tablets by wet granulation. This technology employs a wet binder which provides to the excipient:drug mixture the proper binding characteristics to form granules able to be compressed into tablets. The aim of this study is to evaluate the viscosity, agglomeration behavior and tableting characteristics of cassava starch and to compare these properties to those of classic binders such as alginic acid, sodium alginate, sodium carboxymethyl cellulose, carrageenan, gelatin, polyvinyl pyrrolidone and gum arabic at 5 and 10% levels using ibuprofen as a model drug. Viscosity profiles were determined on a Brookfield viscometer. Wet binders were added to the microcrystalline cellulose-ibuprofen (60:40) mixtures on a ribbon blender and then granules were formed on an oscillating granulator. Granules were then compressed on a single punch tablet machine.

Results indicate that granule properties depended on the binder type, rather than on the binder level employed. Less plastic deforming materials such as alginic acid, gum arabic and PVP k30 rendered tablets with the best compactibility. However, no major differences were observed among the granule properties of cassava and corn starches.

Gum arabic and sodium carboxymethyl cellulose produced the most viscous dispersions (>50000 mPas), which in turn, showed no relation with the strength of the compacts produced. The largest particle size was found for granules made with alginic acid and gum arabic (~290 μm), whereas for most granules it ranged from 100 to 150 μm. The true density of the granules produced ranged from 1.31 to 1.40 g/cm^3. Granules having the largest size also presented the fastest flow rate (>16 g/s). Further, except for alginic acid, sodium alginate and cassava starch, bulk and tap densities ranged from 0.3 to 0.35 and 0.36 to 0.45 g/cm^3, respectively. Moreover, alginic acid and gum arabic produced granules with the lowest porosity (<65%). Likewise, corn starch was the most compressible material (>33%), whereas, alginic acid and gum arabic rendered the least compressible granules. Cassava starch presented the least elastic recovery and along with PVP k30 showed the best reworkability upon reprocessing.

This work demonstrated the potential use of cassava starch as a wet binder for the production of drug compacts by wet granulation.

Keywords: Starch; wet granulation; excipients; wet binders; compaction properties

INTRODUCTION

Starch is one of the most abundant polysaccharides in nature. Significant sources of starch powders come from cassava, corn, rice, wheat, potatoes, etc. Starch is composed of semicrystalline granules of a low water hydration ability (Puchongkavarin et al. 2003). These granules generally have a crystallinity degree from 15 to 45%. Structurally, starch is formed by a tightly packed crystalline lamellae (parallel glucan chains) and amorphous lamellae (less ordered branched chains) (Florence & Roland, 2002).

Cassava and corn starches are dry, odorless, white, tasteless, insoluble and neutral powders, which constitute an essential class of adjuvants used in tablet and capsule formulations. They can be used as a glidant, lubricant, disintegrant or as wet binder in tablet formulations due to their relative low cost, widespread availability, cohesive properties and inertness. These roles depend on the incorporation method and the quantity used (Napaporn et al., 2011).

Compacts are the most prevalent and versatile dosage form. They provide accurate dosing, ease of manufacturing, convenience in administration, and are more stable compared to oral liquids. Further, they are safe compared to injectable dosage forms and tamper resilient as compared to capsules. Typically, compacts can be produced by three technologies: i.e., wet granulation, roller compaction and direct compression. Direct compression is a simple and economic process by which compacts are made directly from a powder blend of a drug and appropriate excipients. In wet granulation, the drug is mixed with a wet binder (usually starch) and other excipients and then passed through sieves to achieve granules which are later dried, milled or sifted before tableting. Wet granulation is still the most popular technology for tableting since it allows a large number of drugs to be compacted in a wide range of doses. The dry granulation process involves the preparation of a dry blend of the drug and excipients, followed by precompression of the powder in high pressure rollers to form ribbons which are then milled and sized. A final addition of a dry binder and/or a lubricant is carried out, if needed, followed by a second compaction step on a tablet press (Rojas, 2012).

Native starches are not employed as a dry filler-binder for direct compression due to their poor flow and compaction properties. Therefore, only pregelatinized starch is used for direct compression of drugs and native starches such as corn and cassava starches are mainly used as wet binders for making granules of excellent flow, compressibility and compactibility (Rojas et al., 2012). The goal of this study is to assess test the functionality of cassava starch as a binder for wet granulation of ibuprofen and compare these properties with commercial binders such as corn starch, gelatin, CMC, alginic acid, sodium alginate, K-carrageenan, gum arabic and PVP k30.

MATERIALS AND METHODS

Materials

Cassava starch (lot 622131), corn starch (lot S0919365) and ibuprofen (lot 2141306132002) were obtained from Famez SA, Grain Processing Corp and Chemo S.A, respectively. Amorphous silicon dioxide (lot I107) was purchased from Cabot Corp. Microcrystalline cellulose PH101 (lot 078/09), sodium starch glycolate (lot 19034/08) and sodium carboxy methyl cellulose grade 4000 (lot 6028/04) were purchased from Blanver. Polyvinyl pyrrolidone k30 (MW 40000, lot P1109011-1) and gum arabic (lot 104K0151) were purchased from Boai NKY Pharmaceuticals and Bell Chem Corp., respectively. Alginic acid (lot 024M0118) and sodium alginate (lot L990972) were obtained from Sigma Aldrich. K-carrageenan (lot 217882) and gelatin (lot 159523) were purchased from Bell Chem Corp. and Protokimica, respectively.

Production of Ibuprofen Granules with Different Binders

Approximately, 100g of ibuprofen and 150g of microcrystalline cellulose PH101 were mixed on a ribbon blender (Riddhi Pharma Machinery, Gulabnagar, India) for five min. Concomitantly, 100 mL of an aqueous dispersion of a wet binder (CMC, gelatin, carrageenan, PVP-k30, corn starch, cassava starch, alginic acid, sodium alginate and gum arabic) was prepared on a blender for 30 seconds to obtain a 5 or 10% w/v concentration. The wet binder dispersion was then added to the previous blend slowly to form a wet mass and mixing was continued for five minutes. The wet mass was then passed through an oscillating granulator equipped with a No. 10 mesh (1680 μm). The wet granules thus formed were then dried on a convection oven (U50, Memmert, Schwabach, Germany) at 105°C for 3h. The dried granules were then mixed with ~5g of sodium starch glycolate (disintegrant) and 0.63 g of amorphous silicon dioxide (lubricant) on a V-blender (Riddhi Pharma Machinery, Gulabnagar, India) for 15 min. The final blend was passed through a granulator equipped with a No. 24 mesh (710 μm).

Granule Properties

Particle size was determined on a Ro-Tap sieve shaker (RX-29, Tyler, Mentor, OH, USA) equipped with size # 325, 200, 140, 120, 100, 80 and 60 screens stacked in the order mentioned. Approximately, 50 g of sample was added to the shaker and operated for 10 min. Particle size was plotted on log-probability scales. The geometric mean diameter (d_g) was calculated from the plots corresponding to 50% cumulative percentage weight using the Minitab software (Minitab, Inc. State College, PA).

The bulk density was determined by taking the ratio of 20 g of sample divided by the measured volume taken on a 100 mL graduated cylinder. The tap density was measured directly from the final volume of the tapped sample obtained from the Auto-Tap analyzer (AT-2, Quantachrome instruments, USA) operated for 300 taps. The true density was determined on a Helium picnometer (AccuPycII 1340, Micromeritics Corp, USA) with ~3 g of sample. Powder porosity (ε) was determined from the equation:

$$\varepsilon = \left[1 - \left(\frac{\rho_{bulk}}{\rho_{true}}\right)\right] * 100\% \qquad (1)$$

where, ρ_{bulk} and ρ_{true} correspond to the bulk and true densities, respectively.

The compressibility of the granules was obtained by applying the Kawakita model (Kawakita & Ludde, 1971):

$$N/C = P/a + 1/ab \qquad C = (V_i - V_n)/V_i \qquad (2)$$

where, N is the tap number, V_i the initial volume and V_n the volume at the respective tap number. The constant "a" is related to the total volume reduction of the powder bed (compressibility index) and the constant "b" is related to the resistant forces (friction/cohesion) to compression.

Flowability, expressed as flow rate (g/s) was determined by filling a glass funnel (13 mm hole diameter) with ~20 g of granules and measuring the flow time through the orifice. Moisture content was determined on an Ohaus moisture analyzer (MB 200, NJ, USA) loaded with ~3g of sample and operated for 5 min.

Preparation of Compacts

Cylindrical compacts of ~500 mg were manufactured on a single station tablet press (Compac 060804, Indemec, Columbia) equipped with a 13-mm flat-faced punches and die tooling at compression pressures ranging from 15 to 140 MPa and a dwell time of 1 s. The upper punch was equipped with a load cell (LCGD-10K, Omega Engineering, Inc, Stanford, CT) and a strain gauge meter (DPiS8-EI, Omega Engineering, Inc, Stanford, CT).

Compressibility Analysis

Compact volume was calculated using the relationship: $V = h\pi r^2$, where *r* (mm) and *h* are the radius and thickness (mm) of the compact, respectively. Compact dimensions (diameter and thickness) were measured with an electronic caliper (Titex Gold, Impofer, 0.01mm sensitivity). Compact densification data were fitted to the Heckel model (Heckel, 1961):

$$Ln(-\varepsilon) = kP+A \quad (3)$$

where, ε is the porosity of the compact at a pressure P and A the intercept, which represents the powder densification by die filling and particle rearrangement before particle deformation and bonding take place. The slope (k) of the linear portion of the plot is inversely related to the mean yield pressure (P_y), which is a measurement of the plasticity of the compressed material. On the other hand, D_a is related to the total compact densification and was obtained from the intercept (A) by the expression:

$$D_a = 1- exp^{-A} \quad (4)$$

Compact Tensile Strength

It was determined on a VanKel hardness tester (UK 200, VanKel, Manasquan, NJ, USA). Each compact was placed between the platens and the crushing force was then measured. The radial tensile strength (TS) values were obtained according to the Fell and Newton (Fell & Newton, 1968) method using the crushing force data. The area under curve of the tensile strength (AUCTS) plots was used to determine the material compactibility. The test was conducted on triplicate.

Compact Disintegration

Cylindrical tablets, each weighing ~500 mg, were made on a single punch tablet press using a 13 mm round, flat-faced punches and die set at a 10% porosity and a dwell time of 1 s. Compacts were tested in distilled water at 37°C employing a disintegration apparatus without discs (39-133-115, Hanson Research Corporation, Northridge, California, USA) operated at 30 strokes/min.

Reprocessing Susceptibility

Round compacts of ~500 g were made on a single punch tablet press using a 13 mm round flat-faced tooling, dwell time of 1 s and 10% porosity. Compacts were analyzed for tensile strength as described under "compact tensile strength" and pulverized on a mortar and pestle. The powder was passed freely on a 24 mesh sieve, compressed again under the same conditions and the resulting tensile strength was determined.

Compact Elastic Recovery

Compacts were made as described under "Reprocessing susceptibility". Once ejected, compact heights were measured with a caliper (Titex Gold, Impofer, 0.01mm sensitivity) and stored in a desiccator (RH = 15-30%), and after five days compacts heights were measured again. The compact elastic recovery was expressed as percentage and calculated from the Armstrong and Haines-Nutt equation as reported previously (Armstrong & Haines, 1972).

Viscosity Studies

Aqueous dispersions of binders (~500 mL) were prepared at 5 and 10% levels on a blender for 30 seconds. Absolute viscosity measurements were conducted on a Brookfield viscometer (RVDVE, Brookfield, Engineer Laboratories INC, Middleboro, MA) equipped with size 61, 62, 63, and 64 spindles. Measurements were conducted at 25°C for 5 min and at 0.6, 1.5, 3.0, 6, 12, 20, 50 and 100 rpm.

Statistical Treatment

A principal component analysis (PCA) was employed. PCA is a type of multivariate analysis that identifies patterns in data and expresses them in such a way as to highlight their similarities and differences in three main axes (PC1, PC2 and PC3). The patterns are the lines that most closely describe the relationships between the data. The PC1 vector is the direction on the abscissa along which projections that have the largest variance. The PC2 is the direction, which maximizes variance among all directions orthogonal to the

PC1. The Minitab® software (v.16, Minitab®, State College, USA) was used for the PCA data analysis. The binder type and binder level (5 and 10%) were taken as independent variables.

RESULTS AND DISCUSSION

Results from the assessment of the ability of cassava starch and commercial binders to form granules of adequate properties are listed on Table 1. In order to guarantee an effective manufacture of compacts from granules, a good binder ideally should render granules with the following characteristics: excellent compressibility, satisfactory flow, resistance to segregation during handling and storage, fast compact disintegration, a high bulk density, and it should be easy to incorporate to the powder mixture during the agglomeration process.

The formation of granules having a good flow is a critical requirement for making compacts with the accurate dose of the active ingredient. Further, a good flow also ensures a uniform die filling during tableting. On the contrary, several problems could come out if granules have a poor flow through the dies of a tablet press. For example, an arch or rat hole may be formed leaving stagnant zones in the hopper. In the best scenario, granules of a large particle size and densities are expected to have the best flowability, especially when a high-speed rotary tablet machine is employed for the compaction process. In this case, alginic acid and gum arabic presented the largest particle size and bulk and tap densities resulting in granules with the fastest flow (>15.9 g/s) and hence, least cohesiveness. This guarantees a homogenous, reproducible and rapid die filling ensuring uniformity in compact weight. Further, the high regularity in shape of alginic acid and gum arabic contributed largely to their good flow (Figure 1). As expected, these two materials had the lowest porosity due to their large packing ability in the powder bed. Conversely, gelatin and carrageenan rendered the smallest granules having a low flowability. It is possible that their irregular shape, and rougher and porous surface texture could induce particle-particle interlocking increasing cohesion and hence, restricting flow.

Table 1. Properties of ibuprofen granules produced from cassava starch and commercial binders

Binder	Level (%)	Geometric mean diameter (μm)	Flow rate (g/s)	True density (g/cm^3)	Bulk density (g/cm^3)	Tap density (g/cm^3)	Moisture content (%)	Porosity (%)
Alginic acid	5	289.3±31.4	17.1±1.3	1.334	0.47	0.54	2.7	65.1
	10	282.4±32.1	16.2±1.2	1.316	0.47	0.54	5.7	64.6
CMC	5	136.4±20.9	12.3±0.4	1.362	0.35	0.45	4.9	74.2
	10	125.3±18.8	9.2± 2.0	1.342	0.34	0.44	3.4	74.3
Carrageenan	5	111.0±16.8	11.9± 0.2	1.376	0.34	0.44	4.9	74.9
	10	123.6±18.6	11.6± 2.2	1.382	0.33	0.43	5.7	75.9
Cassava starch	5	142.3±16.4	15.9± 0.6	1.344	0.37	0.43	3.5	72.5
	10	123.0±12.9	14.5±1.3	1.348	0.36	0.38	3.4	73.3
Corn starch	5	125.9±18.8	13.9± 0.5	1.377	0.33	0.44	4.5	76.0
	10	111.0±16.8	14.4± 0.5	1.388	0.35	0.43	4.8	74.8
Gelatin	5	105.8±15.9	11.5± 0.4	1.372	0.33	0.43	5.3	75.7
	10	108.6±17.7	12.5±2.1	1.402	0.35	0.44	5.3	75.4
Gum arabic	5	297.7±24.9	16.3± 0.8	1.347	0.50	0.57	3.8	62.9
	10	303.5±25.0	15.7± 0.9	1.342	0.48	0.57	2.7	64.5
PVP k30	5	138.7±37.3	14.4 ±0.6	1.311	0.30	0.36	3.5	76.1
	10	131.5±36.2	14.2±0.6	1.317	0.30	0.38	3.3	76.9
Sodium alginate	5	168.7±25.0	12.5 ±0.3	1.354	0.34	0.47	4.5	75.0
	10	173.3±25.8	12.6 ±1.6	1.334	0.34	0.43	4.5	74.2

CMC=Sodium carboxymethyl cellulose, PVP k30=Polyvinyl pyrrolidone.

Table 2. Tableting characteristics of ibuprofen granules produced from cassava starch and commercial binders

Binder	Level (%)	Yield pressure (MPa^{-1})	D_A	AUCTS (MPa^2)	a (%)	b	Elastic recovery (%)	Reprocessing susceptibility (%)
Alginic acid	5	122.0	0.89	187	15.5	0.05	54.7	3.3±3.0
	10	121.1	0.87	225	21.1	0.01	52.6	-9.0±6.1
CMC	5	64.3	0.60	210	26.1	0.13	59.6	32.8±3.1
	10	67.3	0.63	213	21.2	0.10	55.8	22.7±4.1
Carrageenan	5	74.2	0.62	201	21.9	0.03	56.5	17.9±3.6
	10	83.4	0.61	203	26.8	0.03	56.5	19.8±3.6
Cassava starch	5	77.3	0.63	188	25.1	0.05	2.88	-4.5±5.1
	10	83.0	0.66	180	27.8	0.11	2.45	-10.1±3.7
Corn starch	5	78.8	0.61	195	34.3	0.02	56.1	32.9±4.4
	10	77.4	0.59	179	33.1	0.03	54.9	27.9±2.7
Gelatin	5	82.0	0.61	208	22.7	0.07	59.3	23.0±1.6
	10	76.4	0.58	199	22.8	0.04	60.2	32.9±2.3
Gum arabic	5	109.9	0.81	246	13.6	0.02	57.7	-25.3±2.4
	10	126.2	0.82	256	17.5	0.04	63.1	3.4±5.2
PVP k30	5	96.5	0.89	232	14.9	0.06	43.9	-5.7±3.4
	10	107.3	0.88	236	21.5	0.07	56.0	-46.9±4.6
Sodium alginate	5	64.4	0.61	223	29.7	0.03	57.6	15.8±2.1
	10	64.6	0.62	222	22.0	0.06	48.8	16.2±2.1

D_A=Total compact densification, a=Compressibility index, b=Resistant forces to compression, AUCTS=Area under the curve of tensile strength, CMC=Sodium carboxymethyl cellulose, PVP k30=Polyvinyl pyrrolidone.

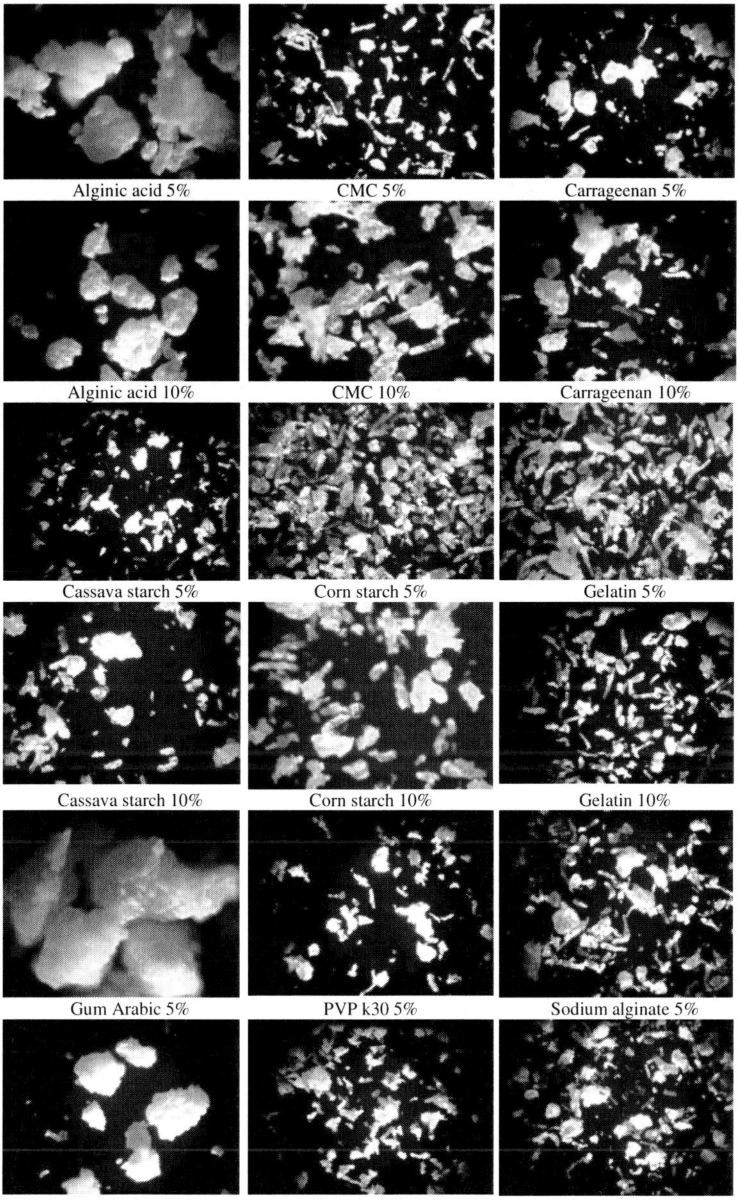

Figure 1. Optical microphotographs of cassava starch and commercial binders taken at 150X magnification.

Granule properties depended on the binder type rather than on the level of binder used. The source of binder also had a large effect on granule properties. For instance, cassava starch rendered granules of larger flowability due to its slightly larger particle size, bulk density and low porosity as compared to corn starch. However, corn starch presented a larger true density. The contribution of moisture can be considered as negligible due to the low water content exhibited by all materials (<5.7%).

Compressibility is expressed as the relative volume reduction of the powder bed in response to the applied pressure, and compactibility is the ability to form a compact with sufficient strength when a compression force is applied. Materials, by virtue of their response to applied forces, can be classified as elastic, plastic, or brittle. Generally, materials are not entirely elastic, plastic or brittle, but have some of all three characteristics. Thus, materials may exhibit all three types of behavior, with one type being the dominant character. Brittleness is caused by progressive failure along weak points in the crystals, whereas ductility favors sliding of crystal planes. A ductile binder can withstand large deformations without breaking, whereas a brittle material fails having a negligible deformation. Ductility and brittleness favor bonding because new surfaces are produced and, consequently, an increase in contact area between particles occurs during compression.

Table 2 lists the tableting properties of granules produced from different binders. The yield pressure (P_y) is related to the type of deformation upon compression. Alginic acid, gum arabic and PVP k30 were the most fragmenting materials upon consolidation in the powder bed. Conversely, all other materials had a more plastic character. Therefore, the yield pressure showed that most materials had P_y < 96.5% except for alginic acid, gum arabic and PVP k30. This indicates that most wet binders increased the plastic behavior of ibuprofen. Further, the ability of corn and cassava starches to render plasticity to ibuprofen granules was comparable and was only exceeded by CMC and sodium alginate. These two materials also rendered the most viscous dispersions at the 5 and 10% levels.

The volume reduction ability of ibuprofen granules expressed as the “a” compressibility index showed a narrow range from 21-35% and was mainly dependent on the particle size. For instance, this index was small for alginic acid and gum arabic, which in turn, exhibited the largest particle size. Gum arabic was the least compressible, whereas corn starch was the most compressible material as indicated by the compressibility index (a). This is explained by the large and smallest particle size exhibited by these two binders, respectively. Further, gum arabic showed more regularly-shaped particles but the small size of corn starch allowed for a better rearrangement and hence, packing easing volume reduction. Further, corn starch presented the least resistant forces to volume reduction promoting compressibility.

Alginic acid, gum arabic and PVP k30 were also the binders that presented the largest total powder densification due to fragmentation and particle rearrangement (D_a). Further, alginic acid, gum arabic and PVP k30 presented the largest particle deformation by die filling. Therefore, when applying a compaction force on these three binders, the crystalline regions could be forced to become closely packed and thus, the intermolecular forces, i.e., van de Waals forces and hydrogen bonding increased, leading to a more ordered rearrangement within the granules. The stronger packing structure resulted in an increase in the tensile strength of the resulting compacts.

The AUCTS (Figure 2a,b) showed two trends in compact tensile strength: one with a fast, and another with a small increase in tensile strength with compression pressure. In the former, gum arabic, alginic acid and PVP k30 reached a plateau in tensile strength sooner than all other materials. Therefore, these materials require low compression forces to achieve a compact with a high tensile strength and thus, those granules yield compacts of sufficient strength without having to apply an excessive compression force. On the contrary, cassava and corn starches, carrageenan, gelatin, CMC and sodium alginate had a steady increase in tensile strength and hence, compression forces of ~65 MPa (for a 5% level) and 80 MPa (for a 10% level) were needed to form compacts of comparable tensile strength to those of alginic acid, gum arabic and PVP k30. Further, cassava and corn starches exhibited a similar compactibility. However, corn starch was more compressible than cassava starch.

The level of binder agent slightly affected the range of compact tensile strength needed to reach a plateau. For instance, PVP k30, alginic acid and gum arabic exhibited a plateau ranging from 1.5 to 2.5 MPa and from 1.8 to 2.5 MPa at the 5 and 10% levels, respectively. Likewise, compactibility, as found by the area under the tensile strength was the largest for the most brittle deforming materials such as gum arabic and PVP k30.

It is important to know the reprocessing behavior of a binder to determine its likelihood for reworking. Sometimes after milling of a reprocessed batch a material could lose its ability to form compacts because the number of binding points is reduced. The reprocessing suscceptibility of binders is expressed as percentage and shown in Table 2. The small or negative change in PVP k30, alginic acid, gum arabic and cassava starch indicates that reprocessing had virtually no effect on the compaction properties of these granules. This phenomenon could be explained by the less plastic deforming ability of these materials. As a result, their crushing force after reprocessing remained almost unchanged. In other words, their low plastic character was not surrendered by reprocessing as compared to other materials which were highly plastic.

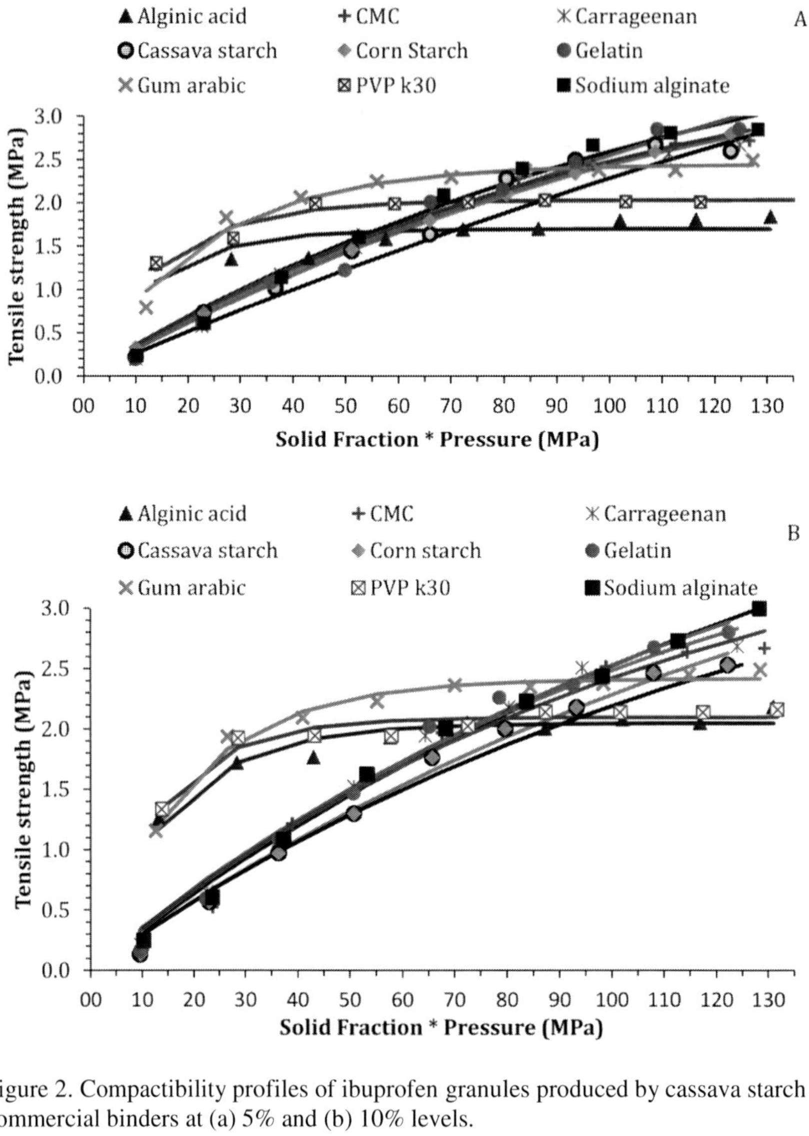

Figure 2. Compactibility profiles of ibuprofen granules produced by cassava starch and commercial binders at (a) 5% and (b) 10% levels.

The elastic recovery (ER) is an important measurement of the robustness of a compact and can be divided mainly into two parts: first, the fast elastic recovery between the maximum compression until detachment of the upper punch from the tablet, and the elastic recovery during compact storage. A fast compact ejection may lead to differences in pressure transmission within the powder bed, preferred orientations in particle bonding and non-homogeneous particle slippage, possibly resulting in tablet lamination and capping. Lamination corresponds to the separation of a compact into two or more distinct horizontal layers, whereas capping occurs when the upper or lower segment of the compact separates from the main body. Table 2 shows that most materials exhibited a large elastic recovery (>43%) upon compact ejection. On the contrary, cassava starch presented the smallest elastic recovery (<3%), which indicates a good behavior after compaction since the applied stored energy was not release upon compact ejection. This could be explained by its low amylose content (14-19% for cassava vs. ~25% for corn starch) which reduced the elastic character upon compression.

In order to understand the overall relationship among all the granule properties a principal component analysis (PCA) was conducted. In this scenario, a loading plot of measured properties depicted in Figure 3 shows projections of these properties onto the PC1, PC2 and PC3 in a 3D plot. The loadings can be understood as weights for each original property when calculating the principal component. This plot is the result of the linear combination of original data that maximizes data variance. Further, each point in the graph indicates the contribution of this property in defining these components. Granule factors contributing very little to the components such as binder level, moisture content and reprocessing showed small loading values and appear plotted near the center. These factors had virtually no effect on the overall behavior of the granule properties, whereas, the binder type was the factor with the largest influence on all properties. Further, compactibility and porosity also appear dispersed around the borders of the plot suggesting these properties are highly affected by the binder type. In this study, the PC1, PC2 and PC3 had a variance of 7.1, 2.7 and 2.2, respectively, and accounted for ~70.5% of the total variance, indicating that most data structure was captured into the three underlying dimensions studied.

The data set related to the densification of granules and deformation upon compression formed a cluster on the left side of the PC1 vector. For instance, particle size was largely correlated with bulk density, flowability and yield pressure. This indicates that binders having a low plasticity are expected to render granules with large particle size, flow rate, bulk and tap densities.

Conversely, granule densities were most directly opposite to the compressibility vectors (i.e., compressibility and porosity) indicating an inverse correlation. Furthermore, some factor pairs such bulk and tap densities, yield pressure and flowability, porosity and forces opposing to compression, viscosity and binder type were also correlated, respectively. On the contrary, the total powder densification was inversely correlated with the compressibility index. Likewise, compactibility was inversely correlated with true density. The resulting scores for PC1, PC2 and PC3 can be expressed as:

$$PC1 = -0.35PS-0.32BD+0.34Por-0.34Py-0.32Da$$
$$PC2 = 0.29Binder-0.41TrD-0.34TD-0.33MC+0.33b$$
$$PC3 = 0.42Binder+0.46Comp+0.53ER+0.35Visc$$

The PC1 vector was related to the densification properties of the granules, whereas PC2 and PC3 were mostly related to the binder factor.

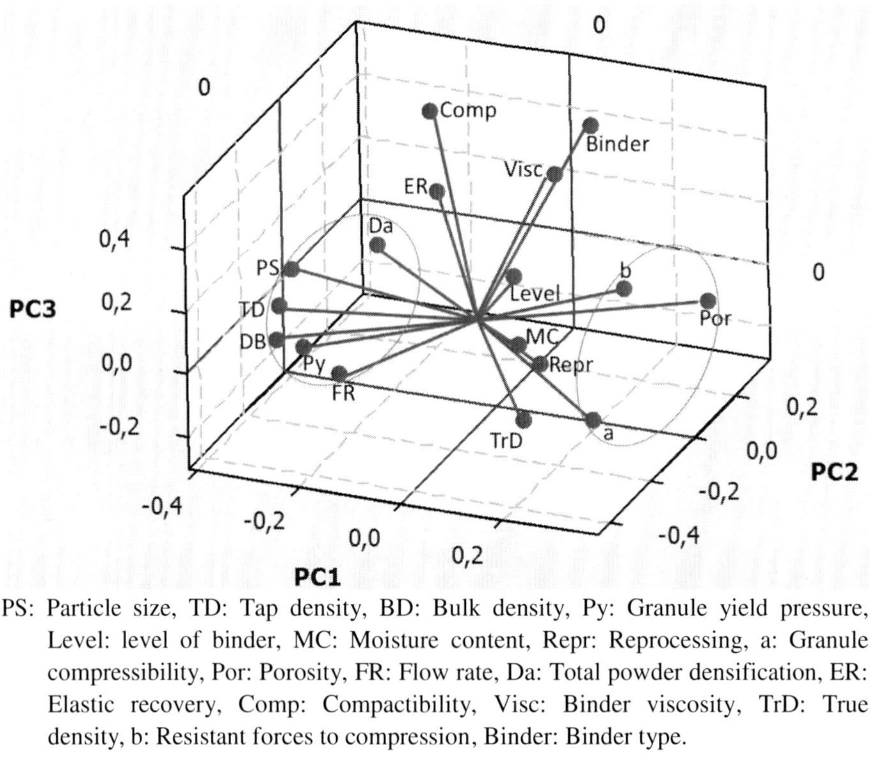

PS: Particle size, TD: Tap density, BD: Bulk density, Py: Granule yield pressure, Level: level of binder, MC: Moisture content, Repr: Reprocessing, a: Granule compressibility, Por: Porosity, FR: Flow rate, Da: Total powder densification, ER: Elastic recovery, Comp: Compactibility, Visc: Binder viscosity, TrD: True density, b: Resistant forces to compression, Binder: Binder type.

Figure 3. Loading plot of cassava starch and commercial binders at 5 and 10% levels.

The 3D score plot of the binders is shown in Figure 4. It separated and classified all binders based on the collective resulting granule properties. For instance, materials such as gum arabic and alginic acid are depicted at the left side of the PC1. These binders were important for producing granules of a large particle size, flowability and densities. These binders also rendered granules with a large powder densification upon compression (D_a), low porosity and large compactibility and packing in the powder bed. On the other hand, binders that rendered granules having a small particle size, densities, flowability, low compactibility, large porosity and reprocessing susceptibility are located at the right side of the PC1. Further, cassava and corn starches appear too separate indicating that granule properties for these two binders were mainly dependent on their source.

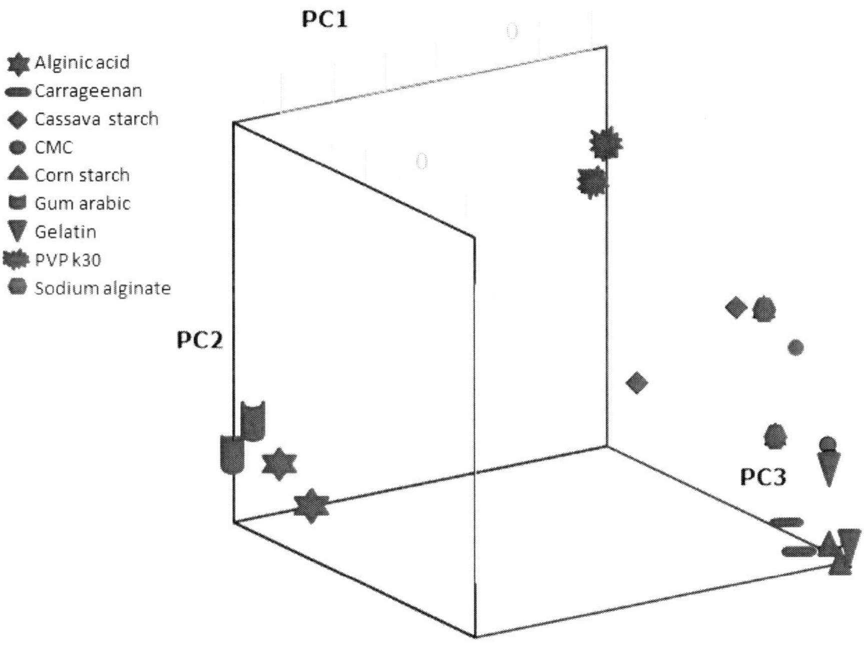

Figure 4. Score plot of cassava starch and commercial binders at 5 and 10% levels.

The disintegration profiles of compacts produced from the binders are shown in Figure 5. Currently, the main acceptable mechanisms of compact disintegration are related to water penetration via water wicking and swelling (Rojas et al., 2012). In this case, the disintegrant (i.e., sodium starch glycolate) functions as a swelling agent which also enhanced the swelling/wicking action of the wet binders. Further, the high hydration ability of cassava and corn starches, CMC, carrageenan and sodium alginate contributed to a fast compact disintegration. In fact, starches presented a high capacity to draw water into the compact since their hydroxyl groups are able to form hydrogen bonds with incoming water molecules. Therefore, starches if incorporated in compact formulations as wet binders, would enhance tablet disintegration by promoting two mechanisms: capillary and swelling. In this case, starches accelerated water uptake and granule separation due to elastic expansion and annihilation of the interparticulate hydrogen bonds. On the other hand, compacts having a low porosity and high compactibility usually have a disrupted water penetration resulting in a sluggish disintegration time (Achor et al., 2010). In this scenario, wet binders that rendered the strongest compacts such as gum arabic, PVP k30 and alginic acid exhibited the slowest disintegrating compacts. On the contrary, all other materials showed a rapid disintegration upon water contact (<500 s). Except for gum arabic and gelatin, the binder level had a negligible contribution on the compact disintegration time. This could be explained by the binder strength which tightened the polymer chains, restricting their swelling ability during the granulation process. Further, resistance to swelling could result from extensive and strongly bonded structures, which is reflected in a greater degree of associative forces in the granules.

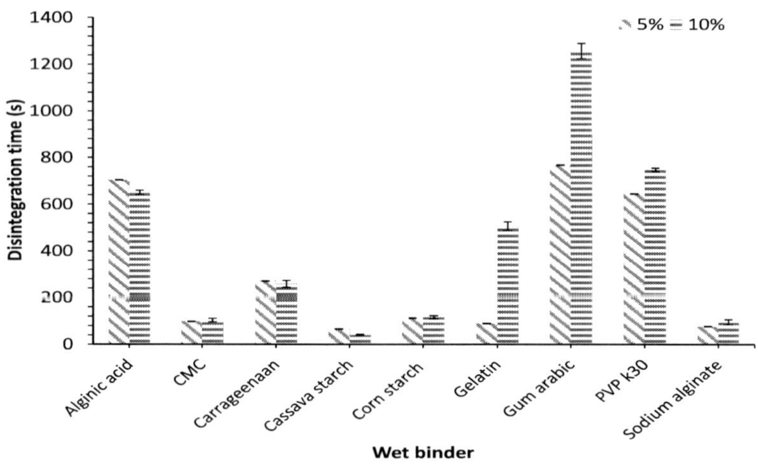

Figure 5. Compact disintegration time of ibuprofen granules produced from cassava starch and commercial binders at 5 and 10% levels.

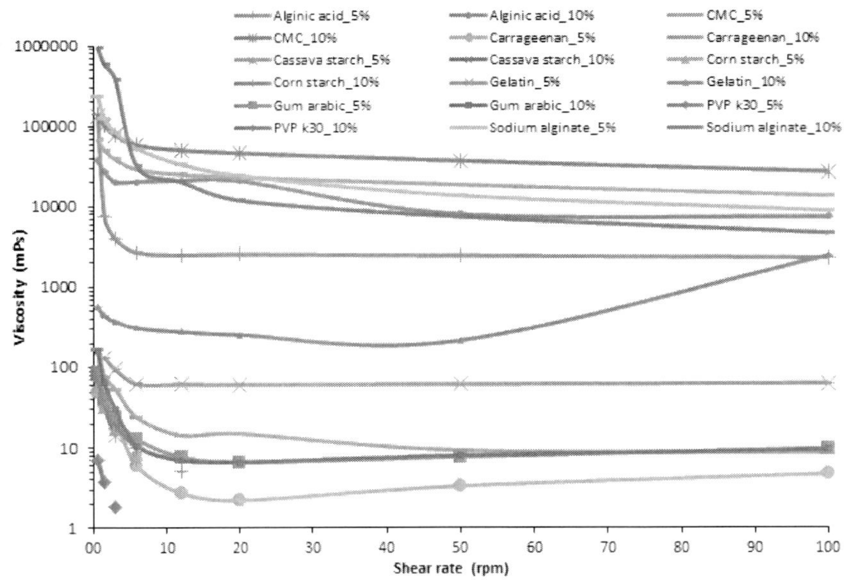

Figure 6. Viscosity curves of cassava starch and commercial binders at 5 and 10% levels and 25°C.

The viscosity curves of the wet binder dispersions are depicted in Figure 6. All binders can be considered as pseudoplastic materials with a yield point equal to a plastic liquid since in most cases, the absolute viscosity decreased sharply and reached a plateau at 10 rpm. This could be attributed to the decreased molecular interactions within of the fluid during flow. Further, additional unfolding and reorientation of polymer chains could be happening. On the other hand, CMC and sodium alginate exhibited a time-dependent viscosity profile. In this case, they exhibited a high viscosity at rest, but a reduced viscosity when submitted to a constant shear stress. Further, in contact with water, CMC, sodium alginate, gelatin and carrageenan formed a strong gel. The intensity of the gelling power depended on the polymer concentration, three-dimensional network of micelles, branching and bonding degrees at the molecular level. It is possible that the high viscosity of sodium alginate and CMC can be attributed to the reorientation and random coiling of the chains once an external stress is applied. It is reported that for starch the gel strength depends on the amylose content, whereas its cohesiveness depends on the amylopectin component (Herman et al. 1989). However, in this study no differences were found between viscosities of cassava and corn starches.

Conclusion

All binders rendered ibuprofen granules that consolidated by particle deformation being alginic acid, gum arabic and PVP k30 the least plastic deforming materials. There was no relationship between the binder viscosity and the resulting strength of the compacts formed. The packing and cohesive properties of the granules depended to a large extent on the particle size distribution and shape, which in turn, depended on the binder type. Thus, granules of small sizes presented a closer packing than large granules made of alginic acid and gum arabic. Further, granules having a low cohesiveness, good flow and compactibility should be the choice when an optimal ibuprofen formulation is required.

Most materials showed similar compaction profiles except for PVP k30, gum arabic and alginic acid. In those cases, compression forces of less than 40 MPa were needed to produce compacts of a high tensile strength. The area under the curve of the tensile strength (AUCTS) indicated that gum arabic and PVP k30 were the most compactable materials. Therefore, it is beneficial to include these two materials as wet binders for making ibuprofen compacts by wet granulation, whereas all other materials presented an intermediate compactibility and starch granules formed compacts with the lowest tensile strength.

Acknowledgments

The authors thank Mr. Edwin Marin of Laproff laboratories for assisting us with the viscosity studies.

References

Achor M, Oyi AR, Isah AB (2010). Tableting properties of acid modified cassava starch dehydrated in alcohol. *CJP Sci.* 4:51-55.

Armstrong NA, Haines-Nutt RF (1972). Elastic recovery and surface area changes in compacted powder systems. *J. Pharm. Pharmacol.* 24:135-136.

Fell JT, Newton JM (1968). The tensile strength of lactose tablets. *J. Pharm. Pharmacol.* 20:657-758.

Florence EE, Roland SO (2002). Effect of acid treatment on the consolidation and plasto-elasticity of tapioca powder. *Trop. J. Pharm. Res.* 1: 45-49.

Heckel RW (1961). An analysis of powder compaction phenomena. *Trans. Metal Sci. AIME.* 221:1001-1008.

Herman J, Remon JP, De Vilder J (1989). Modified starches as hydrophilic matrices for controlled oral delivery. I. Production and characterization of thermally modified starches. *Int. J. Pharma.* 56: 51-63.

Kawakita K, Ludde KH (1971). Some considerations on powder compression equations. *Powder Technol.* 4:61-68.

Napaporn A, Sujin S, Pavinee C, Saiyavit V (2011). A Study of Some Physicochemical Properties of High-Crystalline Tapioca Starch. *Starch.* 53: 577-581.

Puchongkavarin H, Berthaller W, Shobsngob S, Varavinit S (2003) Characterization and utilization of acid-modified rice starches for use in pharmaceutical tablet compression. *Starch.* 55: 464-475.

Rojas J. (2012a). Excipient functionality enhancement: The cellulose II case. The application of the coprocessing technology on excipients. Lambert Academic Press 1: 1-24.

Rojas J, Uribe Y, Zuluaga A (2012b). Material and compaction characteristics of pregelatinized starches. *Pharmazie.* 67: 1-5.

Rojas J, Guisao S, Ruge V (2012c). Functional assessment of four types of disintegrants and their effect on the spironolactone release properties. *AAPS PharmSciTech.* 13: 1054-1062.

In: Cassava
Editor: Francis P. Molinari
ISBN: 978-1-63321-031-8

Chapter 7

Food Products Developed Using Cassava Roots and Its Derivatives: A Review

A. T. Ciarfella[1*], *L. J. Sívoli*[2] *and E. E. Pérez*[3]
[1]Unidad de Cursos Básicos, Núcleo de Anzoátegui, Universidad de Oriente
[2]Departamento de Ciencias Biomédicas, Facultad de Ciencias Veterinarias, Universidad Central de Venezuela
[3]Instituto de Ciencia y Tecnología de Alimentos, Facultad de Ciencias Universidad Central de Venezuela

Abstract

The cassava plant is an important food item in many parts of the world. From this crop may be consumed its protein-rich leaves and its carbohydrate-rich roots with high starch content. However leaves and roots could be potentially toxic to the consumer due to the presence of cyanogenic glycosides, so they must be subjected to effective processing practices for reducing the cyanide in their derivative foodstuffs and these practices depend on the initial level of cyanide. It is of consensus that the roots from varieties called "sweets" have less than 50 ppm of cyanide, while the roots from the varieties called "bitters" have a higher cyanide level, above of 50 ppm. The roots from the "sweets" varieties are boiled or fried just enough to consume but the roots from "bitters" varieties must

[*] Corresponding author: E-mail: anaciarfellaudo@gmail.com.

be crushed or soaked in water for a certain period of time to allow the cyanide volatilization. On the other hand, the fresh roots once harvested have a very short life and the post harvest losses are very high. Is why that the artisan ancient of the local natives and the local food industry have developed technologies of preservation and cyanide elimination, such as: flour elaborated with the "sweets" varieties, gari, farinha or casabe (cassava bread) elaborated with the "bitters" varieties, starch extracted from both varieties and that can be modified to elaborate a myriad of products. With the flour from "sweets" varieties, several products have been developed with or without mixing with wheat flour that have proved to be an alternative for consumers, especially those with dietary special regimes, obesity, cardiovascular disorders, celiac disease or phenylketonurics. Among the products made with 100 % cassava flour are: bread, pasta type "fettuccine", cake and pancake mixes and pizza. With the "catebía" (bitter cassava fermented flour) was prepared "casabe" with high resistant starch content. With the purified starch, native or modified, have been elaborated a kind of crackers and pudding flavored with chocolate. Increasing the consumption of these products, made using cassava roots, helping to reduce wheat flour dependence of the non-cereal producers countries and would be economically advantageous to reduce wheat import levels; this requires changing food habits and to inform the public about the benefits of consuming these products.

Keywords: Cassava flour, food development, *Manihot esculenta*, composite flour, free-gluten flour

The Cassava Plant

The cassava plant (*Manihot esculenta*) belongs to the family *Euphorbiaceae* and is a crop whose origin is assumed to be in the Amazon River Basin (Olson and Shaal, 1999). Depending on the cultivar and environmental conditions can reach between 1-5 m in height and is characterized by a remarkable development of lactiferous vessels that produce a milky secretion (Ceballos and De La Cruz, 2002). Also depending on the plant variety may have palmate leaves with 3 to 7 lobes and a different number of roots (Cartay, 2004). When roots develop, suffer thickening because they store starch and are the part of the plant with the highest economic value (Ceballos and De La Cruz, 2002). The roots are harvested from seven months of planting, depending on the variety, usually at the onset of rains (Cock, 1997). The root yield also depends on the variety and can be between 7-10 t/h in low technology conditions and reach 30 t/h in terms of corporate culture (Cartay, 2004).

Cassava is a tropical crop with extensive agronomic advantages and great climate resilience because it is resistant to drought and efficiently uses the minerals of poor soils (Griffon, 1996; Cartay, 2004). It has high social value because is the main source of carbohydrates for peoples in many tropical countries and has become an agroindustrial culture of extensive development (Montaldo and Montilla, 1996).

What is usually consumed of the cassava plant are their roots, boiled or processed in other products such as flour or starch, although leaves are consumed in some countries of Africa and some populations of the Amazon. Cassava roots are highly perishable so it is necessary to strengthen research on processing and products derived from this crop, to increase market options and provide inexpensive products to consumers (Thro, 1996).

CONCERNS ON THE CYANIDE IN CASSAVA

In the roots, leaves and stems of the cassava plant are stored cyanogenic glycosides (CGs), that to be hydrolyzed by the enzyme linamarase, release a toxic compound for humans and animals, hydrogen cyanide (HCN) a powerful inhibitor of the mitochondrial respiration, in response to cell disintegration. The cassava CGs are linamarin (2-hidroxiisobutironitrilo-β-D-glucopyranoside) and lotaustralin (2-hydroxy-2-methylbutyronitrile-β-D-glucopyranoside), in approximate ratio of 10:1 (Prawat *et al.*, 1995). The release of HCN is an answer to a defense mechanism of the plant against insect attack and also provides repulsive features (Calatayud and Múnera, 2002). The lethal oral dose for humans ranges from 0.5 to 3.5 mg HCN/kg body weight (Badui, 1993). The safe limit of cyanogen in cassava flour has been set at 10 mg CN/kg dry weight (Codex, 1989), which means that an adult with a consumption of 0.5 kg of dry cassava flour per day is exposed to 5 mg of cyanide.

The CGs content in the cassava plant depends on the variety and are present in greater amounts in leaves and bark of the roots (McMahon *et al.*, 1995). On the other hand, the content of CGs in cassava roots determined the use to be given to the production. The Cassava varieties with low cyanide content (<50 ppm HCN) or sweet varieties are sold in markets and are safe for consumers only with being subjected to cooking in water, while varieties with high content of cyanide (> 50 ppm HCN) or bitter varieties require sophisticated methods for processing (drying, soaking, boiling, fermentation) in order that they can be safely consumed. Coursey (1973) proposed a tentative classification based on the reports of several studies; according to this classification, the fresh roots (only the parenchyma) are considered safe if they have ≤50 mg CN/kg fresh weight, moderately toxic if they have between 50 and 100 mg CN/kg wet weight and toxic if they have ≥100 mg CN/kg fresh weight.

There have been cases of fatalities due to acute poisoning by consumption of edible parts of cassava (Padmaja, 1995). Symptoms of cyanide poisoning include palpitations, headache, sore throat, mental confusion, cyanosis, convulsions, coma and eventually death. According to McMahon *et al.* (1995) and Nhassico *et al.* (2008), chronic health disorders associated with a diet high in cyanide include hyperthyroidism, tropical ataxic neuropathy and Konzo. In animals, the lethal doses of HCN are generally reported to be between 0.66 and 15 mg/kg body weight for various species. A sub-lethal dose of dietary cyanide reduced feed consumption and daily growth weight, causes some reproductive effects including resorptions and malformations, lower birth weights and brain weights, increased the number of neonatal deaths, depressed thyroid function, affect the behavioral of the animals (Food Standards Australia New Zealand, 2005).

Consequently, even though cassava is a major source of dietary energy for humans and domestic animals in many countries, its consumption is limited by the characteristic content of CGs.

Concerns by Low Protein Content and Lack of Gluten in Cassava

The cassava plant is a valuable source of carbohydrate, proteins and vitamins but these macro- and micro-nutrients are not well distributed in the plant because cassava roots are rich in carbohydrates but poor in vitamins and proteins, while cassava leaves are an excellent source of protein and vitamins (Montagnac *et al.*, 2009). Because of its low protein content and the presence of cyanide, cassava roots are rated as an inferior food (Padmaja, 1995). In cassava roots about 80% corresponds to the edible portion or parenchyma with 62% moisture, 34% of available carbohydrates, 1% protein, 0.2% fat, 2.3% dietary fiber and 0.6% ash (INN, 2001).

In the comparation of the nutritional composition of cassava with other foods, this no has sufficient protein as cereals or legumes, but has other components as carbohydrate, an interesting option for developing flours. The processing can reduce even more the cassava's nutritional value, especially when the peel is removed. Several biotechnological strategies and research on cassava varieties have been projected to increase protein levels as well as improved essential amino acid content in cassava (Stupak *et al.*, 2006; Boonnop *et al.*, 2009; Gomes and Nassar, 2013).

In animal feeding an alternate practice for incrementing the protein, would be enrich cassava flour with microbial protein (fermentation) (Balagopalan, *et al.*, 1992). Cassava cultivars which are used for cooking purposes have their own leaves rich in protein for enriching itself. For these reasons, the use of cassava leaves as a source of protein serves well as a highly sustainable strategy. Another option to solve this lack, and an easy practice, could be to incorporate flour of cereal and/or legume in cassava flour (Jisha *et al.*, 2008; Duodu and Minnaar, 2011). Continued efforts to improve its nutritional value are important because cassava is a staple food for many people in developing countries.

Moreover, cassava flour, unlike wheat flour does not contain gluten and beyond being a problem could be manipulated to be exploited in the manufacture of foodstuffs for celiac consumers. However, the formulation of gluten-free products using cassava flour is more difficult than containing gluten products, so the preparing remains a major technological challenge (Gallagher, 2008).

FOOD PRODUCTS DEVELOPED USING CASSAVA ROOTS IN VENEZUELA

Cassava is the primary staple for the majority of the Amazonian indigenous people (Lares *et al.*, 2011), also is used in a lot of tropical countries around the world due to the energetic nutritional value that provides to its consumers and has been the third source of calories most important in the tropic after rice and maize (FAO, 2014). Millions of peoples depend on cassava in Africa, Asia and Latin America as staple food as well as for economic income. Its roots highly amylaceous are consumed boiled, roasted and fried, also to produce flour and to isolate starch that have been used to make bread and beverages.

In countries such as Colombia and Brazil is marketed on a large scale with flour and cassava starch. The alcohol from cassava is used in Brazil as a fuel to replace gasoline. The roots can be processed into pellets or pieces to incorporate into formulations for pigs, poultry and fish. The starch is used in the food industry to improve the manufacture, presentation and preservation of food products, hydrolysis products (glucose syrups or maltose, maltodextrin) and isomerization products (isoglucose and fructose) of the starch are used for preparing candies, gels, etc. Starch is also used in non-food industries like textile, paper, pharmaceutical, chemical, etc.

In Venezuela sweet cassava varieties are boiled and eaten with meat, stews and soups or as cassava chips (snack). With bitter varieties is mainly prepared "casabe" (cassava bread), while indigenous populations prepared "mañoco" (grainy), "cachiri" (drink) "katar" (spicy sauce) in addition (Carrizales *et al.*, 1986; Lares *et al.* 2011).

The renewed interest in using cassava for use in another food preparations has its foundation in replace totally or partially the use of wheat flour in the preparations of foods so that they can be consumed by patients with celiac disease, in addition to lowered levels of wheat import because Venezuela is not a cereal producer country.

Flour

The sweet cassava flour is one of the primary products obtained from the roots that has found a wide application in foods. The cassava root flour can be useful for develop food products, emphasizing on gluten free products. In this context Pérez *et al.* (2013) have reported the uses of different root, including cassava for the preparation of different food for special diets. The production of cassava flour is made possible by the use of simple equipment. Pérez *et al.* (2007) have reported the production scheme to elaborate native and modified cassava flours using different thermal treatments that could be used for different baked goods and pasta. These authors reported that the flour preparation procedure affected chemical composition and gelatinization profile.

Pérez and Pérez (2009) have developed a pasta, fettuccine type, using cassava flour (20%) as the wheat flour diluent and beetroot juice. The authors have pointed out that the fettuccine made with this blend have minor protein content as effect of the cassava flour addition, but the cooking time, cooking solids loss and acceptability were similar compared to the control prepared with semolina.

In another studies (Cueto and Pérez 2007, 2008 and 2010; Cueto *et al.*, 2009) has been used cassava flour in order to substitute partially the wheat flour to prepare a dry mix to elaborate cakes. The farinographic profile of the composite flour (30:70, cassava: wheat flours) showed a type I curve, characteristic of flours with low protein content. The cakes prepared with the composite flour had good acceptance, had the characteristics of a conventional cake and the texture profile showed less firmness and chewiness but more cohesiveness and elasticity than the cakes elaborated with100% wheat flour.

Cueto *et al.* (2011, 2012) have developed composite flours (wheat/cassava) for cake and pancake using 30% and 100% of cassava flour. The pancakes 100% wheat flour and 100% cassava flour had presented higher hardness, gumminess and chewiness than pancakes with 30% cassava flour, the uniaxial resistance to extension did not change significantly among pancakes, the substitution of wheat flour by cassava flour modified the internal structure of the products without the occurrence of structure-collapse in the crumb, the products (cakes and pancakes) were well accepted, which are indicative of the feasibility of replacing wheat flour with cassava, and consider the use of 100% formulation cassava as an alternative in the diet for celiac and PKU patients.

Siso and Pérez (2012) formulated a chocolate poundcake by partially substituting the wheat flour with cassava flour and rice bran in a proportion of 50:30:20 (wheat:cassava:rice bran) where the rice bran was used in order to improve the nutritional value. Compared to the control elaborated with 100 % wheat flour, the pouncake showed an increase in the content of ash, crude fat, fiber (insoluble), calcium, phosphorus, and magnesium, the color difference was not significant, the texture profile was slightly different and a high level of acceptance by a semi-trained panel was provided. The moisture content, pH, titratable acidity, texture and fluffiness profiles parameters, measured over three days on refrigerated storage showed consistent stability.

Cassava Bread

The cyanide content in cassava roots has not restricted its use as food for some groups of people who for centuries have developed techniques to detoxify these crop (FAO, 1991). An example of this is that in Venezuela is made "casabe" a kind of circular bread that has been part of the Venezuelan diet for centuries. This cassava bread has about a half centimeter in thick, 60 cm in diameter and 500 g in weight (Lovera, 1996). The cassava bread is cooked on metal plate called "budare" using "catebía", a kind of flour prepared with grated cassava pulp that is squeezed and allowed to ferment for 18 h. The moisture content of the cassava breads is low (10 to 12%) and gives it a great stability in storage and could be without any deterioration for several months (Ciarfella *et al.*, 2013).

The cassava bread is an inexpensive but not nutritious food, although it is an important source of calories for their high starch content. The composition of the cassava bread is: 12.5% moisture, 1.3% protein, 0.6% fat, 80.6% available carbohydrates, 4.1% dietary fiber, 0.9% ash, 60 mg calcium, 78 mg phosphorus, 3.1 mg iron, 0.08 mg thiamin, 0.05 mg riboflavin and 0.7 mg niacin (INN, 2001). As shown it has a low level of protein, so that it is used as a companion of protein sources such as meats and legumes.

The taste of cassava bread is neutral, this product can accept the addition of other ingredients to improve their sensory and nutritional properties. Based on that, a cassava bread fortified with pigeon pea (*Cajanus cajan*) flour was prepared. The legume is of easy access and low cost and the flour was used to increase protein levels. The results obtained showed that with substitution of 25% and 35% of pigeon pea flour, protein levels reached values of 9.46% and 10.88% respectively (Ciarfella *et al.*, 2013). These authors also reported an increase in dietary fiber and resistant starch in cassava breads with legume flour. On the other hand, a lower rate of hydrolysis was also observed in cassava breads with legume flour, suggesting that it could lead to lower postprandial glycemic response. The acceptability of cassava breads containing pigeon pea flour was similar to the control; however some concerns in regards to color and odor were reported by the judges. The increased levels of protein and resistant starch in cassava bread with pigeon pea flour, combined with its lower caloric supply, may benefit people requiring a low carbohydrate diet, such as, diabetic and obese consumers.

Starch

Cassava starch is produced primarily by the wet milling of fresh cassava roots. About 25% starch may be obtained from mature, good quality tubers. The native starch of cassava could be modified through a series of chemical, physical, and enzymatic techniques (Pérez, 1996; González and Pérez, 1996ab; Matos-Segura and Pérez-Sira, 1996; Pérez and González, 1997; Cárdenas *et al.*,

1998; Pérez *et al.*, 1998; González and Pérez, 2003; Matos and Pérez, 2003; Laurentín *et al.*, 2003; Pérez Sira and Pacheco de Delahaye, 2005; Sívoli *et al.*, 2009; Sívoli and Pérez, 2011) and can be used for other industrial applications. Then, a myriad of products could be produced by native and modified cassava starch, such as: bread, canned, beverage, baby food, creams, cheeses, desserts, dips, sausages, etc.

Martinez (1997) prepared crackers with 10% of wheat flour substitution by cassava starch. The crackers were well accepted by semi-trained judges and showed good functional properties.

It was also prepared a recipe of chocolate pudding mix using 21% of cassava starch. The pudding showed similar acceptability compared with commercial one, negative syneresis indicative of a stable gel and good physic-chemical, sensorial and microbiological stability (Andrade, 2007).

On the other hand, the cassava starch has been assayed for use as a matrix of starch-based composite biomaterial for use in biomedical applications, especially tissue engineering (Sívoli *et al.*, 2013). The authors have found that the proposed methylated cassava starch–chitosan system is an excellent matrix that allows cell adhesion and could thereby be further assessed as a scaffold for tissue engineering

Conclusion

The results of these research shows that it is technologically possible diversify the use of cassava derivatives to elaborate several products. The lack of protein is solved using other protein sources such as rice bran and legume flours. By improving technologies and procedures is also feasible the total substitution of wheat flour, offering the celiac consumers gluten free products. The scaling of these products at industry level is suggested, to meet the demand of these consumers due that at today these products are imported and expensive.

References

Andrade, D. (2007). Utilización de almidón de yuca (*Manihot esculenta* Crantz) para la elaboración de una mezcla de pudín saborizado

(Chocolate). Bachelor Thesis of the Facultad de Agronomía, Universidad Central de Venezuela .

Badui, S. W. (2000). Diccionario de Tecnología de Los Alimentos. Ed. Alhambra. México, pp138.

Balagopalan, C., Padmaja, G. & George, M. (1992). Improving the nutritional value of cassava products using microbial technique In:, Proceedings; FAO Animal Production and Health Paper (FAO), no. 95; Expert Consultation on Roots, Tubers, Plantains and Bananas in Animal Feeding, Cali (Colombia), 21-25 Jan 1991 Machin, D.H., Nyvold, S. (eds.). Rome (Italy), FAO. 127-140.

Boonnop, K., Wanapat, M., Nontaso, N. & Wanapat, S. (2009). Enriching nutritive value of cassava root by yeast fermentation. *Scientia Agricola* (Piracicaba, Brazil), *66* (5), 629-633.

Calatayud, P. & Múnera, D. (2002). Defensas naturales de la yuca a las plagas de artrópodos. In: La yuca en el tercer milenio: sistemas modernos de producción, procesamiento, utilización y comercialización. Centro Internacional de Agricultura Tropical. Cali, Colombia. 250-254.

Cárdenas, M., Pérez, E. & Tovar, J. (1998). Pirodextrinización de almidones de yuca y sorgo; un procedimiento para la obtención de material indigerible. In: Memorias del Seminario del Proyecto de Investigación Precompetitiva XI.8. Obtención y Caracterización de Carbohidratos para la Aplicación en Alimentos de Regímenes Especiales. Los Baños, Ecuador. 255-264.

Carrizales V., Barrios O. & González J. (1986). Mejoramiento, tecnológico, de, la producción de cachiri: bebida fermentada de yuca (*Manihot esculenta* Crantz) de origen indígena. *Acta Científica Venezolana*, *37*(3), 318-324.

Cartay., R. (2004). Difusión y comercio de la yuca (M*anihot esculenta*) en Venezuela y en el mundo. Agroalimentaria, n° 18 Enero-junio: 13.

Ceballos, H. & y De La Cruz, G. (2002). Taxonomía y morfología de la yuca. In: La yuca en el tercer milenio: Sistemas modernos de producción, procesamiento, utilización y comercialización. Chapter 2. Centro Internacional de Agricultura Tropical. Cali, Colombia. 16-32.

Ciarfella, A. T., Pérez, E., Tovar, J., Sánchez, T. & Dufour, D. (2013). Efecto de la adición de harina de quinchoncho (*Cajanus cajan*) sobre la calidad nutricional del casabe. *Revista de la Facultad de Agronomía de la Universidad del Zulia*. *30*(1), 131-148.

CODEX. (1989). *Norma del CODEX para la harina de yuca comestible*. CODEX STAN 176-1989.

Coursey, D. (1973). *Cassava as food: toxicity and technology*. In: Chronic cassava toxicity. Proceeding of the Interdisciplinary Workshop (B Nestel y R MacIntyre, eds) London, 27-36.

Cueto, D. & Pérez, E. (2007). Propiedades reológicas de la mezcla lista para torta elaborada con harina compuesta trigo:yuca. Memorias de CIBIA VI. VI Congreso Iberoamericano de Ingeniería en Alimentos. *Ambato, Ecuador*. *16*(1), 133-136.

Cueto, D. & Pérez, E. (2008). Formulación, caracterización y estabilidad de una mezcla lista para torta a base de harina de trigo y yuca. Memorias del V Congreso Internacional. XVI Congreso Nacional de Ingeniería Bioquímica VI. Jornadas Científicas de Biotecnología Molecular y Biomedicina. Chiapas, México.

Cueto, DE., Pérez, E., Borneo, R. & Ribotta, P. (2009). Efecto de la adición de harina de yuca sobre las propiedades reológicas de tortas. *Alimentos Hoy, Revista de la Asociación Colombiana de Ciencia y Tecnología de Alimentos*. Vol *18*. Available in: http://www.acta.org.co.

Cueto, D. & Pérez, E. (2010). Proximate composition and rheological properties of a cake mix elaborated using composite flour wheat:cassava. *International Journal of Food Engineering*, *6*(3), Available in: http://www.bepress.com/ijfe/vol6/iss3/art14 .

Cueto, D., Pérez, E., Ribotta, P. & Borneo, R. (2011). Efecto de la adición de harina de yuca (*Manihot esculenta* Crantz) sobre las características sensoriales, reológicas y físicas de tortas y panquecas. *Revista de la Facultad de Agronomía (UCV)*, *7*(2), 64-74.

Cueto, D. E. & Pérez, E. E. (2012). Propiedades nutricionales de tortas y panquecas a base de yuca y suero de leche bajo en fenilalanina para regímenes especiales. Memorias del Primer Congreso Venezolano de Ciencia, Tecnología e, Innovación en el Marco de LOCTI y del PEI, Caracas, Venezuela.

Duodu, K. G. & Minnaar, A. (2011). Legume composite flours and baked goods: nutritional, functional, sensory, and phytochemical qualities. In: Flour and Breads and their Fortification in Health and Disease Prevention, Chapter 18. Preedy V., Watson R. and Patel V. (eds). 193-203.

FAO. (1991). Cassava bread technology and its future. Food and Agriculture Organization of the United Nations, Roma, Italia.

FAO. (2014). ¿Por qué la mandioca? Available in: http://www.fao.org/ag/agp/agpc/gcds/index_es.html.

Food Standards Australia New Zealand. (2005). *Cyanogenic glycosides in cassava and bamboo shoots*. A human health risk assessment. Technical report Series No. 28, New Zealand.

Gallagher, E. 2008. Formulation and nutritional aspects of gluten-free cereal products and infant foods. Chapter 14. In: Gluten-Free Cereal products and Beverages. *Food Science and Technology*, 321-346.

Gomes, P. T. & Nassar, N. M. (2013). Cassava interspecific hybrids with increased protein content and improved amino acid profiles. *Genetic Molecular Research*, *12*(2), 1214-1222.

González, Z. & Pérez, E. (1996a). Comportamiento amilográfico del almidón de yuca sometido a cocción por extrusión. In: La Yuca Frente al Hambre del Mundo Tropical. A. Montaldo, Compilador. Maracay, UCV, Facultad de Agronomía. 325-33.

González, Z. & Pérez, E. (1996b). Amylographic Performance of cassava starch subjected to extrusion cooking. Chapter 15. In:, Cassava Flour and Starch Progress in Research and Development. Dufour D., O´Brien G.M., Best R. (eds). CIAT-CIRAD. Colombia/ Francia. 128-132.

González, Z. & Pérez, E. (2003). Evaluación fisicoquímica y funcional de almidones de yuca (*Manihot esculenta* Crantz) pregelatinizados y calentados con microondas. *Acta Científica Venezolana*, *54*, 127-137.

Griffon, D. (1996). Adding value to products, byproducts, and waste products of small and medium-scale cassava-processing industries. In: Cassava flour and starch. Progress in research and development. Dufour D., O`Brien G., Best R. (eds). CIAT, Colombia.

INN. (2001). Instituto Nacional de Nutrición. Tabla de Composición de Alimentos para Uso Práctico. Publicación No. 54. Serie Cuadernos Azules; Caracas (Venezuela).

Jisha, S., Padmaja, G., Moorthy, S. & Rajeshkumar, K. (2008). Pre-treatment effect on the nutritional and functional properties of selected cassava-based composite flours., *Innovative Food Science and Emerging Technologies*, *9*(4), 587-592.

Lares, M., Pérez, E., Schroeder, M., Gestne, A., Case, C., Brito, S. & Ciarfella, A. (2011). Biochemical and anthropometric markers, metabolic syndrome and main dietary habits of a Waraos population sample. *Food and Nutrition Sciences*, *2*(5), 444-450 .

Laurentín, A., Cárdenas, M., Ruales, J., Pérez., E. & Tovar., J. (2003). Preparation of indigestible pyrodextrin from different starch sources. *Journal of Agriculture and Food Chemistry*, *51*(18), 5510-5515.

Lovera, J. (1996). Alimentación e historia en Venezuela colonial: El caso de los panes. *Anales venezolanos de nutrición*, *9*, 43-54.

Martínez, R. (1997). Caracterización física-química de galletas tipo soda elaboradas con 3 y 5 % de almidón pre gelatinizado de yuca y harina de trigo. Bachelor Thesis of the Facultad de Agronomía, Universidad Central de Venezuela .

Matos-Segura, M. E. & Pérez-Sira, E. (1996). Comparación de las propiedades del almidón de yuca (*Manihot esculenta* Crantz) fosfatado a escala de laboratorio con uno comercial. *Tecnología de Alimentos, México*, *6*, 23-27.

Matos, M. E. & Pérez, E. (2003). Characterization of native and modified cassava starches by scanning electron microscopy and X-ray diffraction techniques. *Cereal Food World*, *48*(2), 78-81.

McMahon, J., White, W. & Sayre R. (1995). Cyanogenesis in cassava (*Manihot esculenta* Crantz). *Journal of Experimental Botany*, *46*(288), 731- 741.

Montagnac, J. A., Davis, C. R. & Tanumihardjo, S. (2009). Nutritional Value of Cassava for Use as a Staple Food and Recent Advances for Improvement. *Comprehensive Reviews in Food Science and Food Safety*, *8*,191-194.

Montaldo, A. & Montilla, J. (1996). La yuca frente al hambre del mundo tropical. In: La yuca frente al hambre del mundo tropical. Publicación especial de la Universidad Central de Venezuela- Facultad de Agronomía y de Ciencias Veterinarias. Anauco Ediciones, C.A. Caracas, Venezuela. 561 p.

Nhassico, D., Muquingue, H., Cliff, J., Cumbana, A. & y Bradbury, H. (2008). Risisng African Cassava production, diseases due to high cyanide intake and control measures. *Journal of the science of food and agriculture*, *88*, 2043-2049.

Olsen, K. M. & Schaal, B. A. (1999). Evidence on the origin of cassava: Phylogeography of *Manihot esculenta. Proceeding of the National Academy of Science USA*, 96, 5586-5591.

Padmaja, G. (1995). Cyanide detoxification in cassava for food and feed uses. *Critical Reviews in Food Science and Nutrition*, *35*(4), 299-339.

Pérez, E. (1996). Modificación física de almidón de yuca (Manihot esculenta Crantz). In: La yuca frente al hambre del Mundo Tropical. A. Montaldo, Compilador. Maracay, UCV, Facultad de Agronomía. 313-32.

Pérez, E. & González, Z. (1997). Functional properties of cassava (*Manihot esculenta* Crantz) starch modified by physical methods. *Starch/Stärke*, *49*, 49-53.

Pérez, E., Breene, W. & Bahanasey, Y. (1998). Variation in the gelatinization profiles of cassava, sagu and arrowroot native starches as measured by different thermal and mechanical methods. *Starch/Stärke*, *50*, 70-72.

Pérez, Sira, E. E., Pacheco & de Delahaye, E. (2005). Chapter 20. Almidones modificados de raíces y tubérculos tropicales. In: Carbohidratos en Alimentos Regionales Iberoamericanos. Ed. CYTED Programa Iberoamericano de Ciencia y Tecnología para el Desarrollo. Franco M. Lajolo; Elizabeth Wenzel de Menezes (eds). 467- 518.

Pérez, E., Lares, M., González, Z. & Tovar, J. (2007). Production and characterization of cassava (*Manihot esculenta* Crantz) flours using different, thermal treatments. *Interciencia*, *32*(9), 615-619.

Pérez, E. & Pérez, L. (2009). Effect of the addition of cassava flour and beetroot juice on quality of fettuccine. *African Journal of Food Science*, *3*(11), 352-360.

Pérez, E. E., Mahfoud, A., Domínguez, C. L. & Guzmán, R. (2013). Roots, tubers, grains and bananas; flours and starches. Utilization in the development of foods for conventional, celiac and phenylketonuric consumers. *Food Processing & Technology*, *4*(3), 1-6.

Prawat, H., Mahidol, C., Ruchirawat, S., Prawat, U., Tuntiwachwuttikul, P., Tooptakong, U., Taylor, W. C., Pakawatchai, C., Skelton, B. W. & White, A. H. (1995). Cyanogenic and non-cyanogenic glycosides from *Manihot esculenta. Plant Chemistry*, *40*(4), 1167-1173.

Siso, K. & Pérez, E. E. (2012). Uso de harina de yuca *y* salvado estabilizado de arroz en la formulación de panqué enriquecido. Memorias del Primer Congreso Venezolano de Ciencia, Tecnología e, Innovación en el Marco de LOCTI y del PEI, Caracas, Venezuela.

Sívoli, L., Pérez, E. & Gibert, O. (2009). Evaluación del contenido de amilosa del almidón de cassava (*Manihot esculenta* C.) durante la modificación por entrecruzamiento-acetilación empleando el método colorimétrico y de calorimetría de barrido diferencial (DSC). Alimentos Hoy, Revista de la Asociación Colombiana de Ciencia y Tecnología de Alimentos. Vol *18*. Available in: http://www.acta.org.co.

Sívoli, L. & Pérez, E. (2011). Evaluación por fluorescencia de rayos X del contenido de fósforo del almidón de yuca (*Manihot esculenta* C.) durante su modificación por entrecruzamiento., Memorias del VIII Congreso Iberoamericano de Ingeniería de Alimentos. CIBIA 8, Lima Perú.

Sívoli, L., Pérez, E., Caraballo, D., Rodríguez, J. P., Rodríguez, D., Moret, J., Sojo, F., Arvelo, F., Tapia, M., Colina, M. & Barreto-Álvarez, J. (2013). Cytocompatibility of a matrix of methylated cassava starch and chitosan. *Journal of Cellular Plastics*, *49*(6), 507–520.

Stupak, M., Vanderschuren, H., Gruissem, W. & Zhang, P. (2006). Biotechnology approaches to cassava protein improvement. *Trends in Food Science & Technology*, *17*(12), 634-641.

Thro, A., Roca, W. & Henry, G. (1996). The cassava biotechnology network and biotechnologies for improving the processing quality of cassava. In: Cassava flour and starch. Progress in research and development. Dufour D., O´Brien G.M., Best R. (eds). CIAT, Colombia.

In: Cassava
Editor: Francis P. Molinari
ISBN: 978-1-63321-031-8

Chapter 8

PROXIMATE COMPOSITION, PHYSICAL CHARACTERIZATION AND CYANIDE CONTENT OF CASSAVA ROOTS GROWN AT THE GERMOPLASM BANK OF VENEZUELA

A. T. Ciarfella[1]*, L. J. Sívoli[2] and E. E. Pérez[3]

[1]Unidad de Cursos Básicos, Núcleo de Anzoátegui, Universidad de Oriente
[2]Departamento de Ciencias Biomédicas, Facultad de Ciencias Veterinarias, Universidad Central de Venezuela
[3]Instituto de Ciencia y Tecnología de Alimentos, Facultad de Ciencias Universidad Central de Venezuela

ABSTRACT

Cassava is a crop that has spread over a wide area of the American tropics from Venezuela to the northwest of Brazil. It have been established that roots containing less than 50 ppm of hydrogen cyanide are considered sweet, while those with cyanide above this value are bitter. However, it is unknown whether there are differences in the composition of the clones grown in the germplasm bank of the Venezuela country and information on the changes of composition as function of the harvest of cassava; which is between 7 and 11 months after the development of the plant. The goals were to characterize and to compare physical properties,

* Corresponding author: E-mail: anaciarfellaudo@gmail.com.

edible portion, proximate composition and cyanide concentration of four varieties of cassava (*Manihot esculenta* Crantz) from the Germplasm Bank of Venezuela, in different stages of growth. Four varieties coded as: UCV-2192, UCV-2306, UCV-2593, and UCV-2681, were selected. Results of total cyanide classified the varieties UCV-2192 and UCV-2681 such as bitters and it was also observed that the total cyanide content were decreasing as plants was growing, varying from 360 to 118 ppm on the eleven month. The other clones (UCV-2306 and UCV-2593) shown less than 50 ppm of total cyanide and the concentration did not shown any tendency on the growing. The morphology of roots and plants are summarized as: roots and stems midsize well developed. According to these criteria, the sweet clones showed more homogeneous roots and medium size, better developed cylindrical with conical peduncles, than that those of the bitter one. The number of roots by plant was not affected by the cassava, increasing from 6 to 10 units at the last month. The weight of the roots increased significantly from 7 to the 11 months, varying from 169 to 410 g. However, the UCV-2681variety has roots with low weight (165-197g). In regards to the length and diameter of the roots, both parameters have increased with the stage of growing, and the bitter varieties have lower (16.7-17.6 cm) length than the sweet varieties (22.2-26.2 cm). The external color of the roots was lightly high in the bitter varieties, not varying with the growing stage. The edible portion and its color expressed with whiteness index were not affected in all varieties by the growing stage, varying from 70-72 % and 86.5-87.5 % respectively. The edible portion has a tiny elevation in the sweet clones. The content of moisture (63.8-64.5 %), crude protein (0.43-0.53 %), crude fat (0.22-0.25 %), dietary fiber (1.9-2.0 %), ash (0.92-0.96 %), and available carbohydrate (30.7-32.6 %) were not affected by the growing stage. But, moisture and available carbohydrate contents were different with clone type: moisture content was less and available carbohydrate content higher in the sweet clones than in the bitter one. Consequently, the information obtained from this study of the cassava roots and as they are growing will assess their potential and will improve the products and by-products that are produced with them.

Keywords: Cassava roots, cyanide content, *Manihot esculenta*, proximate composition, physical attributes

INTRODUCTION

The cassava crop (*Manihot esculenta* Crantz) was domesticated by South American natives for hundreds of years and subsequently led to West Africa during the colonial period (Cartay, 2004). This plant is also known in other countries as mandioca, aipi, manioc, tapioca, suahili, mhogo and omowgo (Ceballos and De La Cruz, 2002). The cassava plant has the agronomic benefits that can efficiently use the mineral reserves of infertile soils, is resistant to climate changes and long periods of drought (Griffon, 1996), making it a highly efficient carbohydrate producer crop due to its starchy roots, and therefore an important energy source for millions of peoples (O'Brien et al., 1992). The cassava roots are consumed boiled in water, as flour or as other products, such as cassava bread called « casabe » in Venezuela. This kind of cassava bread can remain at room temperature for several months without deterioration by its low moisture (Ciarfella et al., 2013). The production cycle for this crop is in average one year and its roots begin to store starch from 3 months to 9-15 months of planting (Cock, 1997). The most important variables of cultivation and post-harvest use for the cassava roots are: productivity, proximate composition and cyanide content. The production is important because it determines the availability of the crop and specifies the commercial varieties. According to statistics published by the Food and Agriculture Organization of the United Nations (FAOSTAT, 2013) the cassava cultivated area in the worldwide (year 2012) was of 19,990,556 Ha for an average production of 128,325 Hg/Ha, while in Venezuela, as same year, the cultivated area was 36,000 Ha to an average production of 144,444 Hg/Ha, ranking 27th worldwide as a producer of cassava. For the year 2009, in regard to the uses to which cassava production is intended, worldwide 95 million ton were used for human consumption and 432 thousand ton in Venezuela, as animal feed 79 million ton globally and 49 thousand ton in Venezuela. In terms of chemical composition, the roots have (wb) high moisture content, close to 60%, followed by approximately of 35% carbohydrate, mainly as starch, and the remainder 5% arre proteins, fat, ash and other minor components, such as minerals and vitamins (INN, 2001). Cassava roots also contain an anti-nutritional factor such as the cyanogenic glucosides, linamarin and lotaustralin, which can release cyanide (powerful inhibitor of mitochondrial respiration) by enzimatic action of the linamarasa enzyme, also present in the root. The amount of released cyanide is the primary limiting factor for the consumption of the crop, because it has been associated with some types of pathological disorders, acute poisoning and death (Zaninovic 2002 ; Nhassico et al., 2008). The oral lethal dose range for humans is from 0.5 to 3.5 mg HCN/kg of body weight (Badui, 1993) and the Codex (1989) have established that the safe limit of cyanogen in cassava flour is 10 mg CN/kg of dry weight. Colloquially the cassava varieties are classified as "bitter" or "sweet " based on the qualitative assessment of the content of cyanide in them, but Coursey (1973) proposed a tentative classification based on the quantitative content of cyanide of several studies. According to that classification the fresh roots (only the parenchyma) are considered safe if they are on fresh weight ≤50 mg CN/kg, moderately toxic between 50 and 100 mg CN/kg and toxic ≥100 mg CN/kg. Cyanide content submit in the cassava roots determine how they will be processed in order to remove it to safe levels for consumers. The techniques of traditional processing comprise different combinations of drying, soaking, boiling, and fermenting, allowing linamarase interact with the cyanogenic glycosides, leaving that the hydrogen cyanide can escape to the air or been dissolved in water (O'Brien et al., 1992; Padmaja 1995; Bokanga 1995; Edijala et al., 1999; Haque and Bradbury, 2002). Another variable to be taken into account when studying the cassava cultivars is the time to harvest, it would depend on the experience of the producer as a function of the climatic factors or need for sale or processing byproducts. Safo and Osei (1996) established that depending on cultivar or crop age may be differences in vegetative growth and number of roots per plant, in the dry matter and starch content, in starch properties and in the culinary quality of the roots.

As important as studying byproducts manufactured from cassava roots, is to determined the characteristics that differentiate the varieties. Therefore, in this investigation were determined differences in physical characteristics and chemical composition of cassava roots depending on the time of harvest, for four varieties of cassava.

MATERIALS AND METHODS

The root system of 4 varieties of cassava were collected in the Germplasm Bank of the Faculty of Agriculture of the Central University of Venezuela, located in Maracay, Aragua state, located at 10°17' N and 67°3`O, height of 445 meters above sea level and a average of annual rainfall of 982 mm. The samples were identified as UCV-2193 (bitter), UCV-2306 (sweet), UCV-2593 (sweet) and UCV-2681 (bitter) and 2 plants per variety for 5 months were harvested when the culture reached 7, 8, 9, 10 and 11 months of development. The analysis of the samples was started the same day of harvest.

Number and Weight of Roots Harvested

The number and weight of the roots was determined by averaging the values obtained for each plant.

Physical Characteristics

The roots were cleaned to remove adhering soil and were weighed in a balance (Ohaus, model Explorer Pro). The lengths measurements were taken using a ruler with scale in centimeters and the thickness were measurements using a vernier. The roots were peeled manually to separate the edible part of the bark. The calculation of the edible portion was made according to the following equation:

$$Edible\ portion\ (\%) = \frac{weight\ of\ edible\ portion\ (g)}{total\ weight\ of\ the\ root\ (g)} x\, 100$$

Color

The color was determined with a colorimeter (Hunter Lab, model COLORFLEX CX1819) using 10° observer and D65 illuminant. With the obtained values was calculated the whiteness index (WI) (Hsu et al., 2003) for the edible portion and the brown index (BI) (Akissoe´ et al., 2003) for the bark by the following equations:

$$WI = 100 - \sqrt{(100 - L^*)^2 + a^{*2} + b^{*2}}$$

$$BI = 100 - L^*$$

Proximate Composition

The moisture, crude fat, crude protein, ash and dietary fiber contents were determined using the methodology described by AOAC (2012). Available carbohydrates were calculated by difference.

Cyanide

The cyanide extraction was performed by sample homogenization with extraction medium followed by filtration and the linamarase enzyme (EC 3.2.1.21) was obtained from the root cortical parenchyma (Cooke, 1978). For quantitative determination of the total cyanide was used the modified colorimetric method of Essers et al. (1993) and the absorbance was measured at 605 nm.

Statistical Analysis

The analyses were performed in triplicate and expressed as mean ± SD. ANOVA test was used to determine statistical significance and Fisher's least significant difference test (LSD) to establish differences between means. The Statgrafics Plus Program version 4.0 was used for statistical analyzes.

Discussion of Results

Number and Weight of Roots Harvested

The roots morphology of the cassava varieties was different. The roots had elongated shape typical of cassava roots with a larger diameter in the proximal region than in the middle and distal region. The varieties UCV-2306 and UCV-2681 showed more homogeneous and medium-sized, more conical that cylindrical shape and better developed roots that the varieties UCV-2192 and UCV-2593. The average number of harvested roots depending on the time of harvest is show in the Figure 1. The value ranges from 4.5 (UCV-2306 and UCV-2681, month 7) to 11.0 (UCV-2306, month 9). On average, the total number of roots during the 5 months of sampling was lower for bitter cultivars (7) that the sweets (8) but the difference between them was not significant ($p>0.05$). The results agree with those obtained by Safo and Osei (1996) who reported that depending on the time of harvest (between 7 and 13 months) variations were observed in the number of roots between 3.27 and 5.70, and Marin et al. (2008) who reported for 15 cassava varieties that the average number of roots per plant is between 6.0 and 13.3, while Richardson (2013) reported in a study of 6 cassava varieties that the number of roots per plant was between 1.5 and 4.2.

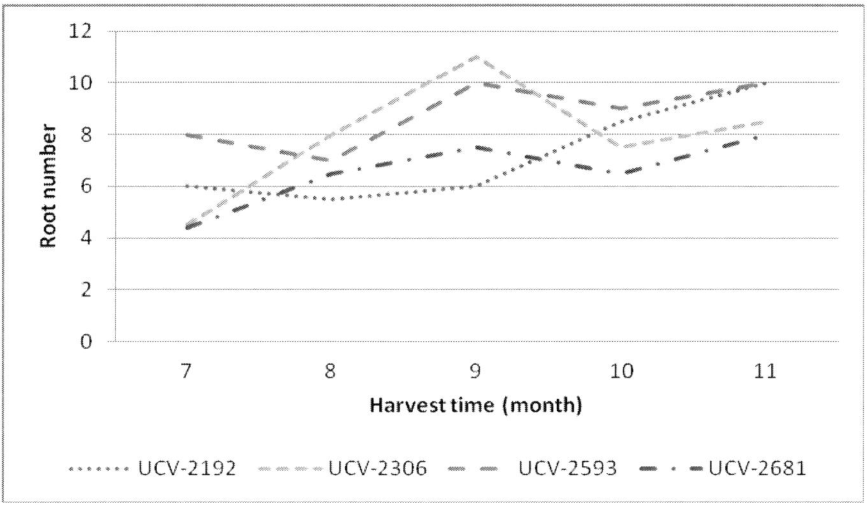

Figure 1. Average number of roots as a function of the harvest time.

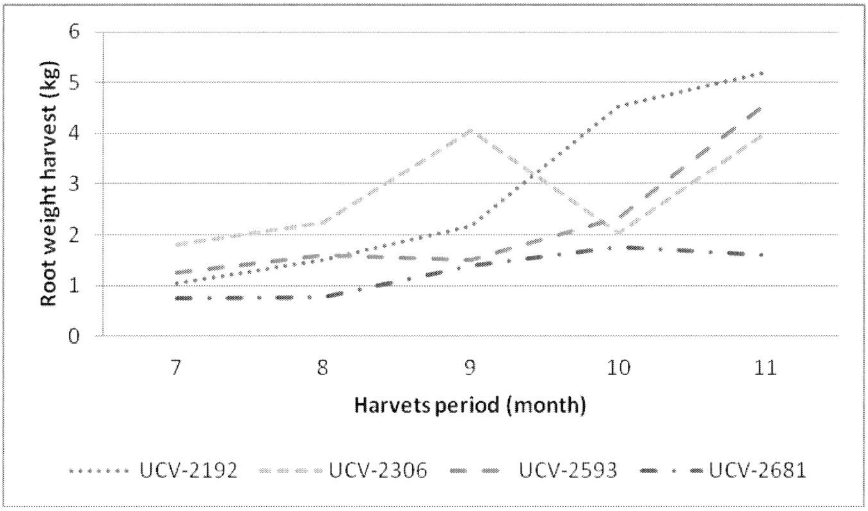

Figure 2. Average weight of the roots harvest as a function of the harvest time.

The roots weight for the cassava varieties is shown in Figure 2. For the UCV- 2192 variety the weight was increased ($p<0.05$) between months 9 and 10 and for the UCV -2593 variety between months 10 and 11, while UCV-2306 and UCV-2681 varieties showed no significant changes with respect to the month of harvest. On average during the 5 months of harvest, UCV -2681 variety had the lowest ($p<0.05$) weight (1.26 kg), the average weight of UCV-2192, UCV-2306 and UCV-2593 was respectively 2.89 kg, 2.83 kg and 2.25 kg. The variety UCV-2192 had the highest average weight and it was the one that had the lowest average number of roots per plant, providing heavier roots. The average weight of cassava roots per plant of the bitter varieties was 2.08 kg and 2.54 kg for the sweet varieties, with no significant difference ($p>0.05$) among them. The values obtained were consistent with those of Richardson (2013) who reported that on a study of 6 varieties of cassava root the weight was between 1.27 and 2.60 kg.

Physical Characteristics

The root length did not differ significantly ($p>0.05$) relative to the sampling month (Figure 3), but on average UCV-2192 and UCV-2681 roots were shorter ($p>0.05$) (17.57 and 16.65 cm) than UCV-2306 and UCV-2593 (22.24 and 26.15 cm). The CODEX (2002) states that the sweet cassava should not be less than 20 cm if it will be sold fresh to the consumer, which meet UCV-2306 and UCV-2593 varieties.

The thickness of UCV-2192 roots varied significantly ($p<0.05$) during the harvest months, from 3.62 cm to 7.17 cm; the roots from the UCV-2306 variety significantly increased ($p<0.05$) thickness between months 10 and 11 until reaching 5.80 cm. The UCV-2593 roots increased ($p<0.05$) thickness from the month 9 until reaching a value of 4.41 cm at month 11. Finally, the roots from UCV-2681 variety had no significant changes ($p<0.05$) of thickness during the harvest months (Figure 4). On average UCV-2593 and UCV-2681 roots were the less thick ($p<0.05$) (3.57 and 3.94 cm), followed by UCV-2306 (4.72 cm) and UCV-2192 (5.64 cm) with the greatest thickness ($p<0.05$). Safo and Osei (1996) have also reported that depending on the time of harvest (between 7 and 13 months) variations were observed in the thickness (0.84 to 3.75 cm) of the roots and the values decline from the month 12.

The average root weight showed no significant changes ($p>0.05$) with respect to the month of harvest, except UCV-2593 variety after the month 10 (Figure 5). On average UCV-2681 roots had the lowest weight ($p<0.05$) (191.07 g) and UCV-2192 roots the highest ($p<0.05$) (391.16 g); UCV-2306 and UCV-2593 roots presented values of 329.30 g and 248.07 g.

The edible portion of the roots did not differ significantly ($p>0.05$) during the harvest months, except for the UCV-2593 variety where the edible portion was increased ($p<0.05$) from the month 10 (Figure 6). On average the less amount ($p<0.05$) of edible portion corresponded to the UCV-2192 variety (68.99%) and the higher amount ($p<0.05$) corresponded to the UCV-2306 variety (73.41%); the UCV-2593 and UCV-2681 varieties had values of 72.19% and 70.94%.

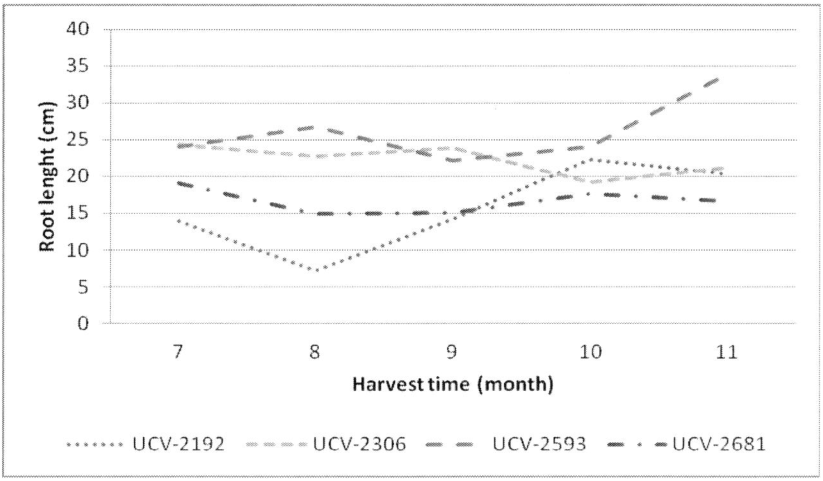

Figure 3. Average lenght of the roots as a function of the harvest time.

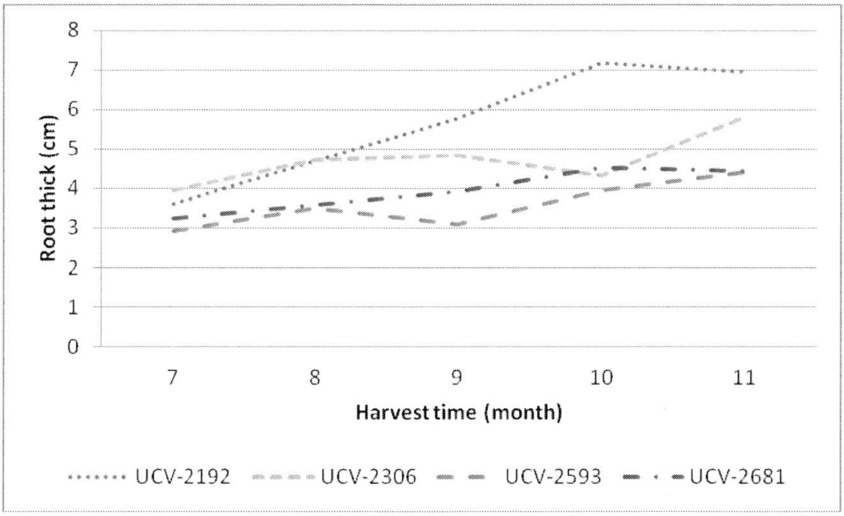

Figure 4. Average thick of the roots as a function of the harvest time.

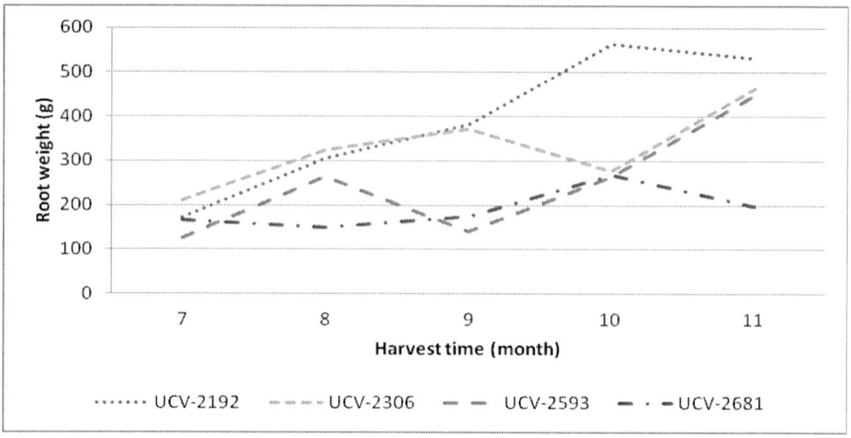

Figure 5. Average weight of the roots as a function of the harvest time.

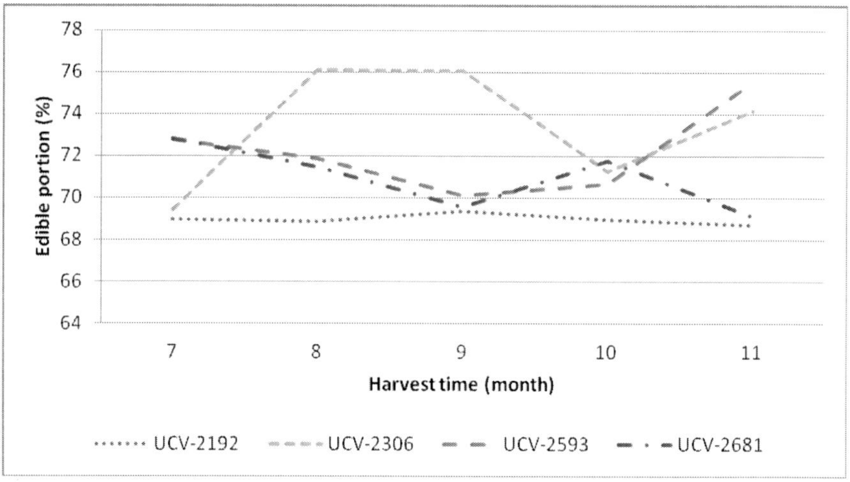

Figure 6. Average edible portion of the roots as a function of the harvest time.

Color

The Table 1 shows the results of color analysis performed to the bark and the edible portion of the roots. The BI of the bark on cassava roots was lower p<0.05) for UCV-2306 variety (51.30) and higher (p<0.05) for UCV-2681 variety (59.15); UCV-2192 and UCV-2593 varieties showed values of 54.57 and 52.43. The bitter cassava varieties have higher value of BI that the sweet cassava. The WI of the pulp was lower ($p<0.05$) for UCV-2593 (85.86) and UCV-2681 varieties (86.46) than for UCV-2192 (87.48) and UCV-2306 varieties (8713). Cassava is a food where the flesh color influences consumer judgment because is expected to be white and any deviation may result in rejection of it, just as happens with products made from these roots, such as flour or cassava where the only ingredient for preparing comes from the grated pulp.

Proximate Composition

The results of proximate composition of the cassava roots are shown in Table 2. During root growth, variations ($p<0.05$) in the moisture content were observed, UCV-2593 variety had the lowest content (58.39%) ($p<0.05$) followed by UCV-2306 variety (62.30%) and bitter varieties (UCV-2192 with 70.19% and UCV-2681 with 68.04%) with the highest content ($p<0.05$). Also Ebah *et al.* (2012) reported that the harvest period (11, 13, 15 and 17 months) and the variety affect the dry matter content, the minimum obtained was 31.17% at 11 months (variety 90/00039) and the highest of 42.99% at 13 months (variety 971A), while Richardson (2013) reported that 6 varieties of cassava had a moisture content between 56.50% and 68.80%. On average, protein values for UCV-2192 (0.43%), UCV-2306 (0.46%) and UCV-2681 (0.43%) varieties had a lower and similar value ($p>0.05$) that UCV-2593 variety (0.55%). Richardson (2013) reported that 6 varieties of cassava had protein content between 1.20% and 2.10%. The fat levels changed ($p<0.05$) during the harvest months but on average no difference were observed ($p>0.05$) among the varieties, the obtained values were 0.18% (UCV-2306), 0.21% (UCV-2192), 0.22% (UCV-2681) and 0.23% (UCV-2593). Richardson (2013) reported in a study of 6 varieties of cassava that crude fat content was between 0.20 and 0.41%. The dietary fiber levels in the roots changed during the five months of harvest, without definite trend, but the average levels of UCV-2593 (2.31 %) were higher ($p<0.05$) than those obtained for UCV-2192 (1.73 %), UCV-2306 (1.86 %) and UCV-2681 (1.85 %) varieties. The National Institute of Nutrition in Venezuela (INN, 2001) reported that dietary fiber content in cassava roots is 2.3 %; Ezeala (1984) reported crude fiber of

0.56 %, Aryee et al. (2006) reported between 0.81% and 2.3 % of crude fiber, Ceballos et al. (2007) reported in two cassava clones 2.6 % and 3.2 % of crude fiber. In the cassava varieties, the ash content was between 0.62 % to 1.20 %, no definite trend observed during the development of the roots, and on average the UCV-2681 variety have the lowest value ($p<0.05$) (0.79 %), followed by the UCV- 2192 (0.86 %) and UCV-2306 (0.88 %) varieties, while UCV-2593 variety had the highest value ($p< 0.05$) (1.09 %). The values obtained were lower than those reported by Richardson (2013) for 6 varieties of cassava (between 2.27 % and 3.24 %). The available carbohydrate levels in the varieties were between 21.77 % and 39.90 % with variations during the growth of the roots but without no definite trend. In average the bitter varieties had lower content, UCV-2192 (26.57 %) and UCV-2681 (28.86 %), that sweets varieties, UCV-2306 (34.32 %) and UCV-2593 (37.43 %). Richardson (2013) reported that in a study of 6 cassava varieties that the total carbohydrate content was between 26.28 % and 39.62 %.

Table 1. Bark brown index (BI) and pulp whiteness index (WI)

Variety	BI (Bark)	WI (Pulp)
UCV-2306 (sweet)	51.30 ± 1.43^{a}	87.13 ± 0.56^{b}
UCV-2593 (sweet)	52.43 ± 1.61^{ab}	85.86 ± 0.31^{a}
UCV-2192 (bitter)	54.57 ± 1.80^{b}	87.48 ± 0.18^{b}
UCV-2681 (bitter)	59.15 ± 1.88^{c}	86.46 ± 0.37^{a}

Values are expressed as mean ± SD. Same letters in the column (a, b, c) are statistically equal to a confidence level of 95% ($\alpha <0.05$).

Table 2. Proximate composition (%) of the edible portion (wet basis)

Varieties	Moisture	Crude Protein	Crude Fat	Dietary Fiber	Ash	Available Carbohydrate
UCV-2306 (sweet)	62.30 ± 0.39^{b}	0.46 ± 0.03^{a}	0.18 ± 0.07^{a}	1.86 ± 0.19^{a}	0.88 ± 0.05^{b}	34.32 ± 1.16^{b}
UCV-2593 (sweet)	58.39 ± 0.22^{a}	0.55 ± 0.06^{b}	0.23 ± 0.06^{a}	2.31 ± 0.31^{b}	1.09 ± 0.10^{c}	37.43 ± 2.58^{c}
UCV-2192 (bitter)	70.19 ± 0.54^{c}	0.43 ± 0.11^{a}	0.21 ± 0.07^{a}	1.73 ± 0.21^{a}	0.86 ± 0.07^{b}	26.57 ± 3.46^{a}
UCV-2681 (bitter)	68.04 ± 0.25^{c}	0.43 ± 0.09^{a}	0.22 ± 0.06^{a}	1.85 ± 0.25^{a}	0.79 ± 0.13^{a}	28.86 ± 4.75^{a}

Values are expressed as mean ± SD. Same letters in the column (a, b, c) are statistically equal to a confidence level of 95% ($\alpha <0.05$).

Table 3. Cyanide content (mg/kg) of the cassava varieties at different harvest period

	Harvest period (month)				
Varieties	**7**	**8**	**9**	**10**	**11**
UCV-2306 (sweet)	12.0	22.1	14.3	10.10	18.30
UCV-2593 (sweet)	9.5	32.4	6.1	15.50	14.20
UCV-2192 (bitter)	278.1	213.6	184.1	150.90	123.40
UCV-2681 (bitter)	360.3	253.2	234.6	216.8	118.2

Values are expressed as mean ± SD. Same letters in the column (a, b, c) are statistically equal to a confidence level of 95% (α <0.05).

Cyanide

The cyanide content (Table 3) in the roots of bitter cassava for the varieties (UCV-2192 and UCV-2681) decreased during the harvest period and was erratic for the sweet cassava varieties (UCV-2306 and UCV-2593). This behavior was also observed by Obigbesan and Fayemi (1976) in a study conducted between months 9 and 12 of age of the plants and Chotineeranat et al. (2006) in a study with roots between 6 and 12 months of age. Between 7 and 11 months of age the UCV-2192 variety the levels of hydrogen cyanide decreased by 56 % (from 278.1 to 123.40 mg/kg) and the reduction in UCV-2681 variety was 67 % (between 360.3 and 118.2 mg/kg). Quiñones et al. (2007) reported for a sample of bitter cassava (UCV-2241 variety) that the total HCN content was 132.73 ppm in the roots with 12 months old. According to the CODEX (2008) the roots of UCV-2192 and UCV-2681 varieties are considered bitters because they contain more than 50 mg/kg of hydrogen cyanide, while the UCV-2306 and UCV-2593 varieties are considered sweets for the content of hydrogen cyanide to below 50 mg/kg.

Conclusion

The study showed that the time of harvest and the variety affects the number and total weight of harvested roots and the diameter and root weight, while the length was not affected. The proximate composition was affected by the variety and harvest time, but the protein levels of the sweet varieties were not affected by vegetative growth. The content of total cyanide in bitter varieties decreases relative to the age of the crop, while in the sweet varieties is observed an erratic behavior.

References

AOAC. (2012). *Official Methods of Analysis, Association of Official Analytical Chemists.* (15th Ed). Arlington, U.S.A.

Akissoe´, N., Hounhouigana, J., Mestres, C, Nago, M. (2003). How blanching and drying affect the colour and functional characteristics of yam (*Dioscorea cayenensisrotundata*) flour. *Food Chemistry*, *82*, 257–264.

Aryee, F., Oduro, I., Ellis, W. & Afuakwa, J. (2006). The physicochemical properties of flour samples from the roots of 31 varieties of cassava. *Food Control*, *17*, 916-922

Badui, S. (1993). *Química de los alimentos.* Ed. Pearson Educación. México. 648p.

Bokanga, M. (1995). Biotechnology and cassava processing in Africa. *Food Technology*. *49* (1), 86- 90

Cartay R. (2004). Difusión y comercio de la yuca (*Manihot esculenta*) en Venezuela y en el mundo. *Agroalimentaria N°*, *18*, 13-22

Ceballos, H. & De La Cruz, G. (2002). Taxonomía y morfología de la yuca. In: B. Ospina and H. Ceballos (Eds). La yuca en el tercer milenio. CIAT. Cali, Colombia. 17-33

Ceballos, H., Sánchez, T., Morante, N., Fregene, M., Dufour, D., Smith, A., Denyer, K., Pérez, J., Calle, F. & Mestres, Ch. (2007. Discovery of an amylose-free starch mutant in cassava (*Manihot esculenta* Crantz). *Journal of Agricultural and Food Chemistry*, *55*, 7469-7476

Chotineeranat, S., Suwansichon, T., Chompreeda, P., Piyachomkwan, K., Vichukit, V., Sriroth, K. & Haruthaithanasan, V. (2006). Effect of Root Ages on the Quality of Low Cyanide Cassava Flour from Kasetsart 50. *Kasetsart Journal (Natural Science.)*, *40*, 694 – 701

Ciarfella, A., Pérez, E., Tovar, J., Sánchez, T. & Dufour, D. (2013). Effect of the addition of pigeon pea (*Cajanus cajan*) flour on the chemical nutritional and sensory quality of cassava bread ("casabe"). *Revista de la facultad de Agrnomía (LUZ)*, *30*, 131-148

Cock, J. (1997). Capítulo1. La yuca: características e importancia. En: La yuca, nuevo potencial para un cultivo tradicional. CIAT, Cali, Colombia. 240 p.

CODEX. (1989). Codex Alimentarius Commission Codex Standard for edible cassava flour. CODEX STAN 176-1989

CODEX. (2002). Codex Alimentarius Commission. Proyecto de Norma del Codex para la yuca (mandioca) dulce. Informe de la décima reunión del Comité del CODEX sobre frutas y hortalizas frescas. Informe N° Alinorm 03/35 (CL2002/27-FFV).

CODEX. (2008). Codex Alimentarius Commission. Comité del CODEX sobre contaminantes de los alimentos. Documento de debate sobre los glucósidos cianogénicos. CX/CF 09/3/11.

Cooke, R. (1978). An enzymatic assay for the total cyanide content of cassava (*Manihot esculenta* Crantz). *Journal of Science and Food Agriculture. 29*, 345- 352

Coursey, D. (1973). Cassava as food: toxicity and technology. In: Chronic cassava toxicity. *Proceeding of the Interdisciplinary Workshop* (B Nestel y R MacIntyre, ed) London 27-36

Ebah-Djedji, B., Dje, K., Zue, B., Zohoun, G. & Amani, N. (2012). Effect of harvest period on starch yield and dry matter content from the tuberous roots of improved cassava (*Manihot esculenta* Crantz) varieties. *Pakistan Journal of Nutrition, 11* (5), 414-418

Edijala, J., Okoh, P. & Anigoro, R. (1999). Chemical assay of cyanide levels of short-time fermented cassava products in the Abraka area of Delta State, Nigeria. *Food Chemistry. 64*, 107- 110.

Essers, S., Bosveld, M., van der Grift, R. & Voragen, A. (1993). Studies on the quantification of specific cyanogens in cassava products and introduction of a new chromogen. *Journal of Science and Food Agriculture. 63*, 287 – 296

Ezeala, D. (1984). Changes in the nutritional quality of fermented cassava tuber meal. *Journal of Agricultural and Food Chemistry, 32*, 467-469

FAOSTAT. (2013). http://faostat.fao.org [January 9/2014]

Griffon, D. (1996). Adding value to products, byproducts, and waste products of small and medium-scale cassava-processing industries. In: Cassava flour and starch. Progress in research and development. Ed. Dufour, O`Brien y Best. CIAT, Colombia

Haque, R. & Bradbury, H. (1999). Preparation of linamarasa solution from cassava latex for use in the cassava cyanide kit. *Food Chemistry. 67*, 305- 309.

Hsu, C-H., Chen, W., Weng,Y-M., Tseng, C-Y. (2003). Chemical composition, physical properties, and antioxidant activities of yam flours as affected by different drying methods. *Food Chemistry, 83,* 85–92.

INN. (2001). Instituto Nacional de Nutrición. Tabla de Composición de Alimentos para Uso Práctico. Publicación No. 54. Serie Cuadernos Azules; Caracas (Venezuela).

Marín, A., Perdomo, D., Albarrán, J., Fuenmayor, F. & Zambrano, C. (2008). Evaluación agronómica, morfológica y bioquímica de clones élites de yuca a partir de vitroplantas. *Interciencia*, *33* (4), 365-371

Nhassico, D., Muquingue, H., Cliff, J., Cumbana, A. & Bradbury, H. (2008). Rising African cassava production, diseases due to high cyanide intake and control measures. *Journal of the Science of Food and Agriculture*, *88*, 2043-2049.

Obigbesan, G. & Fayemi, A. (1976). Investigation on Nigerian root and tuber crops. Influence of nitrogen fertilization on the yield and chemical composition of two cassava cultivar (*Manihot esculenta*). The *Journal of Agricultural Science*, *86*, 401- 406

O'Brien, G., Mbome, L. Taylor, A. & Poulter, N. (1992. Variations in cyanogen content of cassava during village processing in Cameroon. *Food Chemistry*. *44*, 131-136.

Padmaja, G. (1995). Cyanide detoxification in cassava for food and feed uses. *Critical Reviews in Food Science and Nutrition*, *35* (4), 299-339

Quiñónez, R., González, C., Polanco, D., Perdomo, B. & Araque, H. (2007). Evaluación de diferentes tipos de deshidratación de raíz y follaje de yuca amarga (*Manihot esculenta*) sobre su composición química. *Zootecnia Tropical*, *25*(1), 37-41.

Richardson, K. (2013). Quality characteristics, root yield and nutrient composition of six cassava (*Manihot esculenta* Crantz) varieties. Gladstone Road Agricultural Centre Crops Research - Report no. 18. http://www.bahamas.gov.bs. [January 11/2014].

Safo-Kantanka, O. & Osei-Minta, M. (1996). Effects of cultivar and age at harvest on the dry matter, starch gelatinization properties and the cooking quality of cassava. *Ghana Jnl agric Sci*, 28-29, 81-89

Zaninovic V. (2002). Emergencias en salud pública. *Colombia Médica*, *33* (3), 129-130

In: Cassava
Editor: Francis P. Molinari
ISBN: 978-1-63321-031-8

Chapter 9

EXTRA-EARLY MATURING CASSAVA VARIETIES: BREEDING STATUS AND PROGRESS

B. Olasanmi[1,2,*], M. O. Akoroda[1], E. Okogbenin[2], C. Egesi[2] and M. Fregene[3]
[1]Department of Agronomy, University of Ibadan, Nigeria
[2]National Root Crops Research Institute (NRCRI), Umudike, Nigeria
[3]Donald Danforth Plant Science Center, St Louis, MO, US

SUMMARY

This chapter reviews the status and progress made so far in breeding for extra-early maturing cassava (*Manihot esculenta*) varieties. Late root bulking is a major factor leading to rejection and abandoning of improved cassava genotypes in sub-Saharan Africa. Extra-early bulking and high yielding (EB-HY) cassava varieties also referred to as extra-early maturing (EM) varieties in this review shorten the growth period from planting to harvesting. Some EM cassava genotypes have been selected for different agro-ecological zones in Nigeria after evaluation for two seasons at three locations representing three agro-ecological zones. Molecular markers associated with EB-HY in cassava were also identified using Bulked Segregant Analysis (BSA). The selected EM cassava genotypes offer cassava growers high productivity in less than 12

* Corresponding author: E-mail: bunminadeco@yahoo.com.

MAP. This will reduce the problems associated with long dry season experienced in most parts of cassava growing areas as farmers can afford to harvest their cassava plants at the beginning of dry season in areas where rainy season lasts for about 6-7 months. Also, the molecular markers identified for EB-HY in cassava avail breeders the opportunity to select for EM genotypes at early stage of breeding programme and thereby reduce the cost of such breeding programme.

INTRODUCTION

Cassava is the most important of the root crops in the tropics and ranks fourth after rice, sugarcane and maize as a source of calorie for human needs (CIAT, 1992; IITA, 2000). Since its introduction into Africa in the 16th century, cassava has spread throughout sub-Saharan Africa to become one of the dominant starchy staples in the diet of the people. Cassava is a major staple for about half of the Nigerian population and nearly 50 million tonnes of fresh cassava is produced annually in the country (FAOSTAT, 2008). This quantity (about 70% of the production in West Africa) is enough to meet the calorie requirement of 50 million people at about 200 calories per day. Cassava is important not just as a subsistence or food security crop, but also as a main source of cash income for producing households. Its production and processing provide employment and income for the rural poor, especially women and children (Sarma and Kunchai, 1989).

Cassava's long growth cycle makes it relatively difficult for the crop to be readily available on time to farmers and consumers for food and income. Rainfall pattern in some parts of the tropics where rain lasts for about 6 months or less provides distinction between rainy and dry seasons. Late root bulking (LB) is a major factor leading to rejection and abandoning of improved cassava genotypes in sub-Saharan Africa due to demographic and market pressures while early productivity is one of the important traits of interest to farmers in cassava growing regions (Nweke *et al.*, 1994). Early maturing (EM) varieties shorten the growth period from planting to harvesting and better fit into environments with short rainy season, reduce exposure to biotic and abiotic stresses thereby increasing productivity (Nweke et al., 1994). The expanding demands for cassava as food, feed and industrial raw material have made genotypes with high yield and early bulking attributes to be highly desired, hence, improvement of early bulking may have a dual output of early maturing varieties and improved crop yields (Okogbenin and Fregene, 2002). The importance of cassava to humanity in ensuring food security; the ever increasing economic importance of cassava as an industrial crop; and the problem of drought spell in the savanna justify the need to improve cassava for high and early root yield. Extra-early maturing (EM) cassava varieties will ensure shortened production period and thereby solve the problem of drought spell and short rainy season experienced in the dry savanna. Cultivation of the EM varieties will also reduce the period over which farmers wait before they can get some food/income from their farms and thereby improve the role of cassava as a food security crop. In addition, EM cassava varieties may be important to young farmers starting their own farm and requiring early harvest to make efficient use of limited resources such as funds and land. EM cassava varieties can provide such people some food and income few months after harvesting of short duration companion crops included in the mixed cropping which is the prevalent farming system among farmers in Nigeria. Such companion crops (for example, maize, cowpea, groundnut and vegetables) can provide the farmers some food and income for the first 1 to 6 months. They can then start to harvest the EM cassava varieties at about 7-8 months after planting to provide the household additional food and income.

BULKING IN CASSAVA

Bulking refers to the secondary swelling or thickening of the storage roots as they are filled with excess assimilates after the plant might have satisfied the needs for vegetative growth and respiration. The cyclical accumulation of dry matter in storage roots of cassava is responsible for the increase in yield experienced over the growth period. Phasic development experienced in cassava implies a sequence of four growth phases for a crop harvested after 12 months, and five growth phases for a crop harvested after 18 months (Osiru *et al.*, 1995). Cassava continues the first bulking until about 9 MAP and then moves to the recovery stage which lasts till about 15 MAP; the second bulking starts around 15 MAP and this phase continues till about 19 MAP (Osiru et al., 1995). Wholey and Cock (1974) stated that bulking commences in cassava after 2 months of growth while Okogbenin and Fregene (2002) reported starch initiation as early as 6 weeks after planting (WAP) and rapid differentiation in root size at 9 WAP. Root yield is directly influenced by the rate and duration of bulking with highest yielding genotypes having the highest rate of bulking for the longest period of time (Okogbenin and Fregene, 2002). EM cassava varieties start bulking early and bulk more rapidly than the late bulking varieties, hence, the EM variety (A) gives higher yield than the late bulking one (C) at early stage of development (Figure 1). All the three varieties in the Figure are high yielding but the late bulking variety leaves it late before rapid bulking. Variety B is an intermediate between late and early bulking varieties.

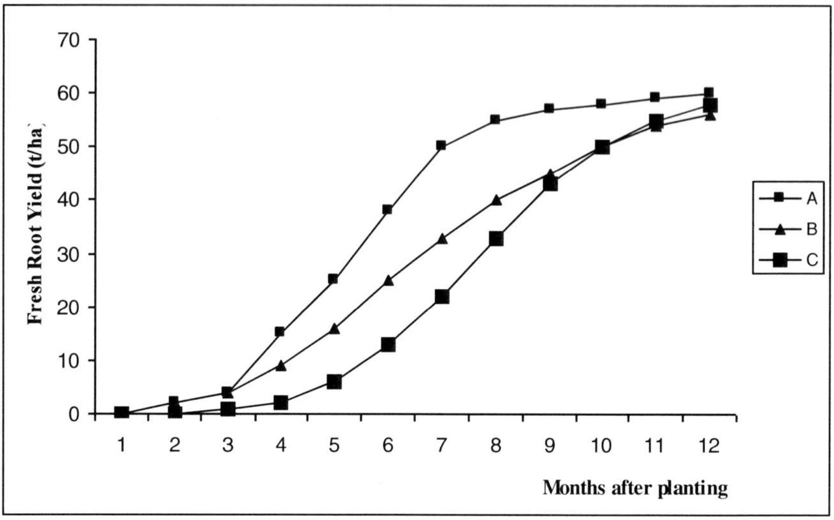

Figure 1. Hypothetical illustration of bulking rate in three cassava varieties.

CASSAVA GERMPLASM RESOURCES

Crop breeding can be successfully carried out when there is an ample availability of germplasm resources and thorough understanding of such available germplasm resources. Cassava germplasm is characterised by a wide genetic diversity which is used by breeders to improve cassava yield, quality and adaptability to biotic and abiotic stresses (Hershey and Jennings, 1992; El-Sharkawy, 1993). A cassava germplasm containing nine hybrid populations is currently maintained by Department of Agronomy, University of Ibadan, Nigeria. Each of the nine populations was developed from hybridization between an early bulking (EB) parent and a late bulking (LB) parent at National Root Crops Research Institute (NRCRI), Umudike, Nigeria. The genotypes can be evaluated further to improve cassava for EB-HY. However, there is need to make more crosses in order to increase the number of progeny in each population and thereby increase the possibility of identifying more molecular markers associated with EB-HY in cassava (Olasanmi, 2010).

DEVELOPMENT OF EXTRA-EARLY MATURING CASSAVA VARIETIES

Nine cassava populations (COB-1 to COB-9) were developed using six elite varieties from the Nigerian cassava germplasm as parents in 2005 (Table 1). The three female parents have been earlier classified as early bulking varieties and the three male parents as late bulking varieties at NRCRI, Umudike, Nigeria. This was further confirmed after evaluation of the varieties for early bulking in two seasons at NRCRI during the course of evaluation of the progeny for early bulking and high yield in 2006 and 2007 (Olasanmi, 2010). The progeny were evaluated for fresh root yield (FRY), fresh shoot weight (FSW), number of storage roots (NSR) and harvest index (HI) at seedling (SET), clonal (CET) and preliminary (PET) evaluation trial stages of breeding programme seven months after planting (MAP) to identify early bulking genotypes referred to as extra-early maturing genotypes in this review.

Thirty-two EM genotypes were selected after PET for multilocational trial (MLT). The selected genotypes and two checks (TMS 30572 and TMS 98/0505) were assessed for FRY, FSW, NSR, and HI at Umudike, Otobi and Ibadan (all in Nigeria) at 7 MAP in randomized complete blocks with three replications in 2008 and 2009. TMS 30572 is currently the national check while TMS 98/0505 is one of the newly released high yielding and cassava mosaic disease (CMD) resistant varieties in Nigeria. White (1983) classified Ibadan to be along the transitional zone between lowland rainforest and derived savanna while IITA (2009) classified the site to be in the derived savanna. Otobi is in the derived savanna while Umudike is in the humid forest (IITA, 2009).

Number of seeds per female flower is genotype dependent in cassava; therefore, selection of highly fertile genotypes as female parents is a critical factor (Kawano, 1980). Maternal effect and compatibility between male and female parents have influence on seed set in cassava (Olasanmi, 2010). Therefore, it is necessary to investigate seed set capability of male and female parents used in the development of recombinant seeds in cassava breeding programmes. This will help the plant breeder to choose parental varieties with high seed set capability and thereby maximize his/her resources (time, labour and funds) (Olasanmi, 2010).

Table 1. Nine cassava hybrid populations developed from six parents at National Root Crops Research Institute, Umudike

Population	Female parent	Male parent	Number of progeny
COB-1	TMS 30572	TMS 30555	181
COB-2	TMS 30572	NR 8212	119
COB-3	TMS 30572	NR 8083	124
COB-4	TMS 97/2205	TMS 30555	148
COB-5	TMS 97/2205	NR 8212	133
COB-6	TMS 97/2205	NR 8083	101
COB-7	TMS 98/0505	TMS 30555	272
COB-8	TMS 98/0505	NR 8212	199
COB-9	TMS 98/0505	NR 8083	217

There was segregation among the progeny of each cross between an early bulking (EB) and a late bulking (LB) parents for EB-HY and associated traits as evidenced by the high variation and significant differences observed among the progeny in each F_1 population for FRY and other related variables during seedling, clonal and preliminary evaluation trials. The desirable transgressive segregants with respect to all the measured and evaluated variables identified in each F_1 population provided the opportunity for improvement of cassava for early bulking and the associated traits.

Significant correlation was observed for FRY, HI and NSR at different stages of evaluation of the progeny. This indicates that the traits were fairly stable across the evaluation stages and may therefore be selected for at early stage of a breeding programme with high precision. This is in support of the findings from earlier studies that correlation of traits associated with EB in cassava is very high between PYT/UYT and CYT (Kawano et al., 1998). Also, the correlation coefficients among yield and yield-related variables are almost the same at SET, CET and PET stages, hence, a breeder may decide to select for the traits at SET stage. However, the genotypes were evaluated with only one plant and few plants at seedling and clonal evaluation stages respectively. About 50% of the genotypes selected as EM at the clonal evaluation stage gave very low yield when evaluated in replicated trial in two seasons at Umudike and were therefore not selected for multilocational trial. Therefore, it is better to delay selection for yield in a conventional breeding programme until advanced stages when the genotypes are evaluated in replicated trials with equal plot size. Significant higher mean DRY at 7 MAP at Otobi than Ibadan and Umudike suggests that EM cassava genotypes performed better at Otobi than the other two locations (Figure 2).

According to El-Sharkawy and Cock (1990), many clones of cassava have harvest indices of 0.6 or more which suggests that there is little scope for breeding for higher yield through improvement in harvest index, hence, increase in yield will have to come from increasing primary productivity. However, the low average HI recorded during the evaluation of the selected EM cassava genotypes at the three locations (0.45 at Umudike; 0.50 at Ibadan and 0.40 at Otobi) suggests that the genotypes can be further improved for higher yield through increased HI.

USE OF MARKER-ASSISTED SELECTION IN BREEDING FOR EARLY MATURING CASSAVA VARIETIES

Marker-Assisted Selection (MAS) is much more important in selecting for traits that are difficult to evaluate or which are expressed late in the growth cycle. Thus, MAS should significantly reduce the costs associated with breeding schemes frequently used in cassava improvement programmes (Okogbenin, 2004). Also, DNA and other molecular techniques have proven useful for the determination of genetic relationships in plants (Kresovich et al., 1992). A statistical methodology for MAS combines information on molecular genetic markers with standard phenotypic measurements to construct optimal selection indices (Lande, 1991).

The ability to identify genetic components of traits, particularly quantitative traits, in Mendelian factors, and to monitor or direct their changes during breeding through the use of DNA-based markers has created much enthusiasm. Claims were sometimes made that MAS would rapidly replace phenotypic selection and dramatically reduce the time required to develop commercial varieties (Mazur, 1995). The exaggerated promises about the power of MAS to improve complex traits created excessively high and largely unfulfilled hopes and prompted a wave of cautious and sometimes pessimistic views (Young, 1999; Goodman and Carson, 2000; Bernardo, 2001). While biotechnology cannot replace conventional breeding methods, integration of the two methods has the potential to facilitate improvement in crop yield and quality.

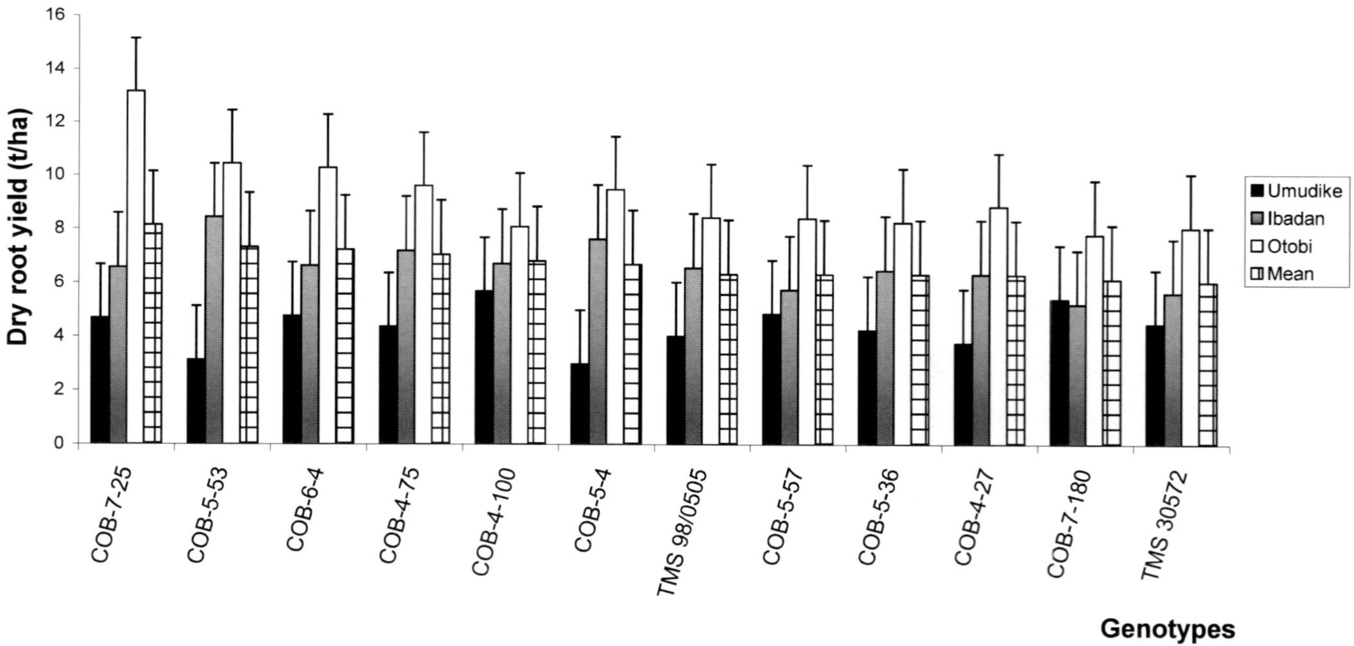

Figure 2. Mean dry root yield (t/ha) of selected early bulking cassava genotypes and two checks evaluated at three locations in Nigeria 7 months after planting.

One of the tools used to identify molecular markers associated with a trait in an organism is **b**ulked segregant analysis (BSA). It measures the variation present in pools of segregants that have been sorted according to phenotype and uses the correlation between these measurements and the pool phenotype to assign a likely map location (Brauer et al., 2006). Some 542 simple sequence repeat (SSR) markers were used to screen the nine cassava hybrid populations (COB 1 to 9) for molecular markers associated with early bulking and high yield (EB-HY) in cassava at the Cassava Genetics Laboratory of International Center for Tropical Agriculture (CIAT), Cali, Colombia in 2009. The 542 SSR markers cover the entire cassava genome and thereby enabled the tracking of a good number of QTLs associated with early bulking in cassava. A high number of SSR markers were found to be polymorphic between 2 parents of each population (ranging from 93 to 131) and thus increased the probability of detecting polymorphic markers between the 2 contrasting bulks of the population (Olasanmi et al., 2014)

Two SSR markers were found to be closely associated with EB-HY in each of populations COB3, COB6, and COB8 while one SSR marker was found to be closely associated with the trait in each of populations COB1, COB7, and COB9. None of the SSR markers was found to be linked with QTLs associated with EB-HY in the remaining three populations (Olasanmi et al., 2014). This may be due to the fact that the cassava genetic map is still unsaturated leading to poor linkage of certain markers (Okogbenin, 2004). Further saturation of cassava genome with more markers may lead to detection of QTLs linked with EB-HY and associated traits in the three populations as well as detection of more QTLs in the remaining six populations. Seven of the nine QTLs to which the markers are linked explained for 10% or more of the phenotypic variation indicating that the seven markers are linked to major QTLs associated with EB in cassava (Olasanmi et al., 2014).

All the five SSR markers found on the existing cassava genomic map (a genetic linkage map of a cassava backcross based on a B_1P_2 family and SSR markers) (Akinbo, 2008) are in the same region of QTLs identified by Okogbenin (2004) in his study. The different locations of the identified QTLs (as indicated by the linkage groups on which the associated SSR markers are located) on the cassava genetic map is indicative of the polygenic nature of most of the traits studied (Okogbenin, 2004).

About half of the genotypes selected conventionally (using phenotypic data) as EM genotypes and evaluated at different locations in Nigeria were selected as EM genotypes by the SSR markers found to be associated with EB-HY and associated traits in cassava. This therefore indicates that use of SSR markers can help breeders select with reasonable precision for desired traits in cassava (Olasanmi et al., 2014).

PLANTING MATERIAL FOR PROPAGATION

Tropical root crops such as cassava and sweet potato are usually propagated vegetatively and African farmers use their seedlings only rarely (Gibson et al., 2000; Manu-Aduening et al., 2005), preferring the more certain option of vegetative propagation from known cultivars. Good planting material is important for high root yield (CIAT, 1984) and the planting material should be healthy (disease-free and pest-free). Quality of stake is determined by plant age, stake size, number of nodes, duration of storage, nutrient status of the stakes as well as differences ascribable to the specific genotypes used (Lozano et al., 1977). For cassava cultivation, stake length of 25 cm (20 – 30 cm) is recommended (Ghosh et al., 1988) and 5-7 nodes per stake may ensure good plant establishment because few buds on stakes do not favour plant establishment (CIAT, 1975). The multiplication rate of cassava using vegetative propagation is very low and this could represent a severe constraint when a new variety is being introduced to a region, when the area under cassava is being increased or when loss of planting material occurs as a result of poor management or other reasons (Henry, 1991). Harvesting of cassava towards the end of rainy season at about 7 to 8 MAP in an environment with distinct rainy and dry seasons may constitute problem in terms of planting material for propagation at the beginning of next rainy season. The farmer cannot establish a new field at the beginning of dry season unless under irrigation and cassava stems can hardly be stored for more than 2 weeks under natural condition before drying up. Also, storage of stems for months in a conditioned environment during the dry season may not be practicable for cassava production due to the magnitude of work involved. There is therefore need to develop a strategy that will ensure availability of enough planting materials during the following planting season for the farmer to establish a new field. The strategy being proposed here to solve this problem is termed "Tithing in cassava production". This strategy requires that each farmer reserves about ten percent of his/her farmland for stem multiplication. Such portion of the field can be planted at a spacing of 1 x 0.5 m to give plant population of 20,000 plants/ha. The farmer can harvest this portion of the field when ready to establish new field the following season and use the stems for propagation. The farmer can still get enough storage roots at harvesting of the multiplication plot to pay for the cost of production of the planting material. However, there is need to test this proposal with some farmers and perfect the strategy.

Conclusion and Future Perspectives

The extra-early maturing cassava genotypes with higher FRY and DRY than the checks offer cassava growers high productivity 7-8 MAP. EM cassava genotypes will ensure shortened production period and thereby solve the problem of drought spell and short rainy season experienced in the dry savanna. This is important for cassava regarding the recent global climatic changes. Cultivation of the EM cassava genotypes will also reduce the period over which farmers wait before they can get some food or income from their farms and thereby improve the role of cassava as a food security crop. In addition, extra-early maturing variety may be important to young farmers starting their own farm and requiring early harvest. There is need to demonstrate the early maturity trait to farmers through participatory on-farm evaluations across the country. The identified materials provide a useful source of early maturity genes for multi-trait pyramiding of useful genes.

References

Akinbo, O.A. 2008. Introgression of high protein and pest resistance genes from inter-specific hybrids of *Manihot esculenta* ssp *flabellifolia* into cassava (*Manihot esculenta* Crantz). *Thesis*. Plant Science, Natural and Agricultural Sciences. University of the Free State, Bloemfontein. 263.

Bernado, R. 2001. What if we knew all the genes for a quantitative trait in hybrid crops? *Crop Sci*. 41:1-4.

Brauer M. J., Christianson, C. M. Pai, D.A. and Dunham, M.J. 2006. Mapping Novel Traits by Array-Assisted Bulk Segregant Analysis in *Saccharomyces cerevisiae*. *Genetics* 173: 1813-1816.

CIAT. 1975. Centro Internacional de Agricultura Tropical. Cassava Production Systems. In: Annual Report for 1974. Cali, Colombia.

CIAT. 1984. Selection and preparation of cassava cuttings for planting. Study Guide. 1984.

CIAT. 1992. Cassava Program Annual Report for 1987-1991. Cali, Colombia: CIAT.

El-Sharkawy, M.A. 1993. Drought tolerant cassava for Africa, Asia, and Latin America: breeding project work to stabilize productivity without increasing pressures on limited natural resources. *Bioscience* 43: 441-451.

El-Sharkawy, M.A. and Cock, J.H. 1990. Photosynthesis of cassava (*Manihot esculenta*). *Experimental Agriculture* 26: 325-340.

FAOSTAT, 2008. Statistical database of the Food and Agricultural Organisation of the United Nations. Retrieved 26th January, 2011 from http://faostat.fao.org. FAO, Rome.

Ghosh, S.P., Ramanujam, T., Jos, J.S., Moorthy, S.N. and Nair, R.G. 1988. *Tuber Crops*. New Delhi: Oxford and IBH Publishing Co. PVT. Ltd.

Gibson, R.W., Jeremiah, S.C., Aritua, V. Msabaha, R.P., Mpembe, I. and Ndunguru, J. 2000. Sweet potato virus disease in Sub-Saharan Africa: evidence that neglect of seedlings in the traditional farming system hinder the development of superior resistant landraces. *Journal of Phytopathology* 148: 441-447.

Goodman, M.M. and Carson, M.L. 2000. Reality vs myth: corn breeding, exotics, and genetic engineering. Proc. 55th Ann. Corn and Sorghum Res. Conf., pp. 149-172. Washington, D.C., USA, American Seed Trade Association.

Henry, G. 1991. Adoption of cassava technologies: constraints, strategies and impact. Ed. G. Henry. *Trends in CIAT Commodities 1991.* Cali, Colombia: CIAT. Document No. 93: 22-40.

Hershey, C.H. and Jennings, D.L. 1992. Progress in breeding cassava for adaptation to stress. *Plant Breeding Abstracts* 62: 823-831.

IITA. 2000. Starting a cassava farm: IPM Field Guide for Extension Agents. 4-10.

IITA. 2009. The Agro-ecological map of Nigeria.

Kawano, K. 1980. Cassava. *Hybridization of Crop Plants*. Eds. W.R. Fehr and H.H. Hadley. American Society of Agronomy and Crop Science Society of America. Madison, Wisconsin, USA. 225-233.

Kawano, K., Narintaraporn, K., Narintaraporn, P., Sarakarn, S., Limsila, A., Limsila J., Suparhan, D., Sarawat, V. and Watananonta, W. 1998. Yield improvement in a multistage breeding program for cassava. *Crop Sci* 38.2: 325-332.

Kresovich, S., Williams, J.G.K., McFerson, J.R., Routman, E.J. and Schaal, B.A. 1992. Characterization of genetic identities and relationships of

Brassica oleracea L. via a random amplified polymorphic DNA assay. *Theoretical and Applied Genetics* 85: 190-196.

Lande, R. 1991. Marker-assisted selection in relation to traditional method of plant breeding. *Proceedings of the Symposium on Plant Breeding in the 1990s*. 437-451.

Lozano, J.C., Toro, J.C.J., Castro, A. and Bellotti, A.C. 1977. Production of cassava planting material. CIAT Series GE – 17. Cali, Colombia. 28 pp.

Manu-Aduening, J.A., Lambll, R.I., Dankyi, A.A. and Gibson, R.W. 2005. Cassava diversity and evolution in Ghanaian traditional farming systems. *Euphytica* 144: 331-340.

Mazur, B.J. 1995. Commercializing the products of plant biotechnology. *Trends Biotech.* 13: 319-323.

Nweke F.I., Dixon, A.G.O., Asiedu, R. and Folayan, S.A. 1994. Cassava variety needs of farmers and potential for production growth in Africa. *Collaborative study of cassava in Africa (COSCA)*. Working paper 10: 40-41.

Okogbenin E. 2004. Quantitative trait loci mapping of root quality traits, morphological characters, and early bulking in cassava (*Manihot esculenta* Crantz). Thesis, Agronomy, Agriculture and Forestry. University of Ibadan. 220.

Okogbenin, E. and Fregene, M. 2002. An SSR-based molecular genetic map of cassava. *Euphytica* 147: 58-66.

Olasanmi, B. 2010. Marker-Assisted Selection (MAS) for improvement of traits associated with high and early root productivity in cassava (*Manihot esculenta* Crantz). Thesis, Agronomy, Agriculture and Forestry. University of Ibadan. xxi +168.

Olasanmi B., M.O. Akoroda, E. Okogbenin, C. Egesi, S.S. Kahya, O. Onyegbule, F. Ewa, J. Guitierrez, H. Ceballos, J. Tohme and M. Fregene (2014). Bulked segregant analysis identifies molecular markers associated with early bulking in cassava (*Manihot esculenta* Crantz). *Euphytica* 195: 235–244 doi: 10.1007/s10681-013-0991-2.

Osiru, D.S.O., Porto, M.C.M. and. Ekanayake, I.J. 1995. Physiology of cassava. IITA Research Guide 55. Training Program, International Institute of Tropical Agriculture (IITA), Ibadan, Nigeria. 7-8.

Sarma, J.S. and Kunchai, D. 1989. Trends and prospects for cassava in the Third World. *Summary proceedings of a workshop on trends and prospects of cassava in the Third World.* Washington D.C.: IFPRI. 20-35.

White, F. 1983. The vegetation of Africa, a descriptive memoir to accompany the UNESCO/AETFAT/UNSO vegetation map of Africa. UNESCO, *Natural Resour. Res*. 20: 1-356.

Wholey, D.W. and Cock, J.H. 1974. Onset and rate of root bulking in cassava. *Experimental Agriculture* 10: 193-198.

Young, N.D. 1999. A cautiously optimistic vision for marker-assisted breeding. *Molecular Breeding* 5: 505-510.

INDEX

#

A

B

C

E

F

G

H

I

K

L

M

N

O

P

Q

R

S

T

U

V

W

X

Y

Z